Historical Reenactment

Reenactment History

Series Editors: Vanessa Agnew, Jonathan Lamb and Iain McCalman

Reenactment has been generally associated with cultural heritage events and living history performances and was seen mainly as the preserve of history buffs and fantasy role players. During the past decade, however, reenactment has emerged as a vital trend in popular as well as scholarly forms of historical representation. It has begun to make its way into historiography as a new concept in the understanding of the past. The Palgrave *Reenactment History* addresses this new interest and embraces ways of understanding diverse experiments in simulative and affective forms of historical representation. It is hospitable to the study of any practical engagement with the past in which the empirical outcome is determined not by what is known in advance, but by the experience of making it.

Common to all these forms of historical representation is an attention to the details of physical, emotional, and psychological experience rather than the sweep of large-scale historical processes, or structures. In one respect reenactment returns the specific historical moment to the discipline of scientific experiment, where what was known once may be known again by means of replicating the original circumstances of the discovery. In another respect it introduces history to passions generated when an individual dwelling in a present moment of time achieves a sympathetic identification with another inhabiting an earlier one. A hypothetical structure becomes a real position. Conjecturalism of this exact and yet powerfully dramatic kind lends reenactment a great emancipatory potential, perhaps in some respects a dangerous one. In ways that other forms of historical representation cannot, reenactment binds the circumstances of the original event to its repetition in order to release energies not usually associated with the garnering of knowledge, and powerful enough to prevent the outcome being predictable. If reenactment is an inquiry whose success depends on the mutuality of cognitive and affective impulses of the mind, implying a new understanding of the task of the reenactor-historian and the very purpose of historical representation, then a critique apt for its novel blend of epistemology and aesthetics is an urgent challenge.

The series aims then to survey new kinds of histories by breaking with traditional approaches. Its volumes treat various historical periods, sites, and disciplinary approaches to historical representation—colonialism, realism, affective cognition, literature, music, art, architecture, drama, performance, agriculture and the sea. With its comprehensive scope and its theoretical edge, the series will interrogate different disciplinary approaches to reanimating and representing the past. In so doing, the editors expect that it will cause us to not only examine what we know about the past, but also how we know it. Finally, it will prompt us to inquire into the social and political uses made of historical knowledge in the present.

Vanessa Agnew and Jonathan Lamb (*editors*)
SETTLER AND CREOLE REENACTMENT

Iain McCalman and Paul A. Pickering (*editors*)
HISTORICAL REENACTMENT
From Realism to the Affective Turn

Reenactment History
Series Standing Order ISBN 978–0–230–20286–3 (Hardback) 978–0–230–20287–0 (Paperback)
(outside North America only)

You can receive future titles in this series as they are published by placing a standing order. Please contact your bookseller or, in case of difficulty, write to us at the address below with your name and address, the title of the series and the ISBN quoted above.

Customer Services Department, Macmillan Distribution Ltd, Houndmills, Basingstoke, Hampshire RG21 6XS, England

---

Historical Reenactment

From Realism to the Affective Turn

Edited By

Iain McCalman

and

Paul A. Pickering

palgrave
macmillan

First published 2010 by
PALGRAVE MACMILLAN

Palgrave Macmillan in the UK is an imprint of Macmillan Publishers Limited, registered in England, company number 785998, of Houndmills, Basingstoke, Hampshire RG21 6XS.

Palgrave Macmillan in the US is a division of St Martin's Press LLC, 175 Fifth Avenue, New York, NY 10010.

Palgrave Macmillan is the global academic imprint of the above companies and has companies and representatives throughout the world.

Palgrave® and Macmillan® are registered trademarks in the United States, the United Kingdom, Europe and other countries

ISBN 978-0-230-57612-4 hardback

This book is printed on paper suitable for recycling and made from fully managed and sustained forest sources. Logging, pulping and manufacturing processes are expected to conform to the environmental regulations of the country of origin.

A catalogue record for this book is available from the British Library.

A catalog record for this book is available from the Library of Congress.

10 9 8 7 6 5 4 3 2 1
19 18 17 16 15 14 13 12 11 10

Printed and bound in Great Britain by
CPI Antony Rowe, Chippenham and Eastbourne

Contents

List of Figures

Acknowledgements

This volume (and the Series of which it is a part) emerged from string of conferences, seminars, and workshops held variously in Nashville, Pasadena, Canberra, Melbourne, Sydney, and Cambridge. Our thanks are due to those others who (in no particular order) presented papers and participated in conversations at and beyond the conference table: Andrea Bradley, Alex Cook, Jed Buchwald, Harry Liebersohn, Jonathan Lamb, Vanessa Agnew, Annie Hesp, David Gruber, Paul Turnbull, Bill Deverell, Robert Rosenstone, Kim McKenzie, Dan Turner, Brian Deyo, Caroline Turner, Jonathan Mane-Wheoki, Dadang Christanto, Vincent Caretta, Cassandra Pybus, Pat Hoffie, Julie Gough, Michael Mel, Adam Low, Toby Haggith, Tatiana Senkevitch, Ian Howard, Claire Roberts, Geremie Barmé, Mark Seltzer, Salima Hashmi, Helena Hammond, Michael Desmond, Anthony White, Helen Weinstein, Patrick Wildgust, James Chandler, Geoffrey Day, Madeleine Bunting, Ashley Hay, Beau Yancy Baca, Gordon Lang, Lynne Blundell, Jennifer Webb, Susan Steggall, Paula Hamilton, Lucinda Mathieson, Chris Linke, Yannick Thorval, Christine Hansen, Amanda Johnson, Tina Parolin, Vivienne Kelly, Robyn Ballinger, Michael Farrell, Otto Sibum, Jim Secord, Alison Winter, Jacques Rancière, Mark Phillips, Elizabeth Edwards, Kim Sloan, Stephanie Lindsay-Thompson, Dominique Sweeney, Will Davies, James Hurst, Melissa McEwen, Ivo Lovric, Anthony Garnaut, Travis Cutler, Mary Kilcane Cody, Anna Garretson, Laurence Gourievidis, Olwyn Beazley, Veronica Tuckerova, Roy Ritchie, Simon Baker, John Wyver, Jim Spriggs and Christopher Pinney.

Thanks also due to those who helped organise the moving feast – Leena Messina, Sally Kingsley and Sabrina Boschetti – and to those institutions which provided funding support: The Australian National University, the University of Sydney, California Institute of Technology, Vanderbilt University, Cambridge University, University of York, and the Huntington Library.

We wish to acknowledge the following institution and individuals for permission to reproduce images: Kanien'kehaka Onkwawen:na Raotitiohkwa Language and Cultural Center, Kahnawake Mohawk Nation Territory, Canada, Stephen Gapps and Dan Hallett (copyright owned by Dan Hallett and Jonathan Walker).

Our sincere thanks to Michael Strang and Ruth Ireland at Palgrave Macmillan for their interest in a new subject and for their forbearance in the face of our tardiness.

Individual authors have included words of thanks, where appropriate, in their chapters. It is left to us to thank them for their efforts in bringing this volume to completion. Working in an emerging area such as reenactment studies carries risks as well as rewards. We are grateful to those willing to take the risk.

Iain McCalman and Paul Pickering

Notes on Contributors

Kate Bowan is a Visiting Fellow at the Research School of Humanities in The Australian National University. She has published on early twentieth-century Australian music, and music and politics in the nineteenth century.

John Brewer is the Eli and Eyde Broad Professor in Humanities and Social Sciences at the California Institute of Technology in Pasadena, California. His most recent book, published this year, is *The American Leonardo: A Twentieth-Century Tale of Obsession, Art and Money*.

Simon During teaches English at Johns Hopkins University. He is theauthor of *Modern Enchantments: The Cultural Power of Secular Magic* and *Exit Capitalism: Literary Culture, Theory andPost-Secular Modernity*.

Stephen Gapps currently works as a consultant historian and is co-director of the history events management company, HISTORICA. His current research interests lie in the problems and possibilities of historical reenactment and public commemoration. Stephen's doctoral thesis 'Performing the Past: A Cultural History of Historical Reenactments' (University of Technology, Sydney) looked at the long history of historical reenactments and its relationship to contemporary practices of self-styled 're-enactors'.

Katie Kitamura was born in 1979 and brought up in California and Japan. She now shuttles between London and New York. She was educated at Princeton University (A.B.) and University of London (MRes, Ph.D.). She has written for numerous publications, including *The New York Times* and *The Guardian*. Her recent books include *Japanese for Travellers* and *The Longshot*.

Iain McCalman is currently Professorial Research Professor at the University of Sydney. His recent books include *The Seven Ordeals of Count Cagliostro* and *Darwin's Armada: How Four Voyagers to Australasia Won the Battle for Evolution and Changed the World*.

Ruth B. Phillips holds a Canada Research Chair in Modern Culture and isProfessor of Art History at Carleton University in Ottawa, Canada. Herresearch focuses on critical museology and the indigenous arts ofnorth-eastern North America. Her books include *Trading Identities: The Souvenir in Native North American Arts from the Northeast, 1700–1900,* and,most recently, *Sensible Objects: Colonialism, Museums and Material Culture,*co-edited with Elizabeth Edwards and Chris Gosden.

Paul Pickering is Professor of History and Director of Graduate Studies at the Research School of Humanities in The Australian National University. His recent books include *Unrespectable Radicals: Popular Politics in the Age of Reform* and *Feargus O'Connor: A Political Life.*

Anja Schwarz is a researcher and lecturer in the literature department of the University of Konstanz. She holds an M.A. in Cultural Studies from the University of Lueneburg and a Ph.D. in English Literature from the Freie Universitaet Berlin. She has published on multiculturalism and historical reenactments as well as questions of gender and memory.

Jonathan Walker is Senior Research Fellow at the University of Sydney and the author of *Pistols! Treason! Murder!: The Rise and Fall of a Master Spy* and *Five Wounds*. The latter is a novel created in collaboration with the artist Dan Hallett. Walker's website is www.jonathanwalkervenice.com.

James Walvin is Professor of History Emeritus at the University of York and has written extensively on slavery and the slave trade.

Trudy Nicks is a senior curator in the Department of World Cultures at the Royal Ontario Museum, Toronto and an adjunct professor in the Anthropology Department, McMaster University, Hamilton, Ontario. She has recently completed work for two new galleries based on the ROM's worldwide ethnographic collections. Her recent publications include papers on early-twentieth-century Iroquoian collectors. Her current research is focused on government intervention in aboriginal arts and crafts production in twentieth-century northeastern North America.

1
From Realism to the Affective Turn: An Agenda

Iain McCalman and Paul A. Pickering

Ever since the late eighteenth century when new forms of visual entertainment claimed to be able to use technology to replicate or simulate the literal details of nature, scholars, poets and intellectuals have bemoaned the increasing dominance of the realist aesthetic. Classicists such as Sir Joshua Reynolds and Romantics such as Samuel Coleridge found the practice of literalist simulation to be debased and disgusting. It demanded merely mechanical competence, eliminated the vital creative role of the artist's idealising imagination, and appealed to uncultivated tastes attracted by the sensational wonders of the technical facsimile. Art lost its age-old *raison d'être* if it became indistinguishable from literal nature.

From the outset, too, these forms of realist technology were linked closely to practices of historical reenactment. The advent of naturalist technologies in theatre scenography almost immediately led to mechanical and visual reenactments of real-life historical events, such as social pageants, battle-scenes and shipwrecks. These in turn quickly passed into the new spectacular visual media of panoramas and dioramas and into cinema thereafter. For many practitioners (and commentators) modern manifestations of reenactment constitute the *ne plus ultra* of realism. Every weekend countless re-enactors enjoy an 'outlet from hum-drum life', taking the 'unique freedom to roam the centuries – to explore the intricacies of our ancestors [*sic*] minds and habits'.[1] Among re-enactors are 'hard core' for whom authenticity is a byword, a means to narrow the gap between past and present, as if counting the stitches on a tunic or soaking leather in urine is a way to experience what it felt like to live in the past.[2]

To modern semioticians such as Roland Barthes, intoxication with the 'reality effect' is quite simply the dominant aesthetic, or more properly anti-aesthetic, of late modernity. Inventions like the commercial panorama are said to have presaged the contemporary fetish for surrogate visual experience, for digital and cinematic special effects, and for the 'hyper-realistic' theme parks, fleshy waxworks and technology-filled museums which so bewilder, bemuse and infuriate commentators such as Umberto Eco. To the French social theorist Jean Baudrillard this contemporary appetite for realistic simulation leads ultimately to the enthusiastic consumption of the simulacrum, a real-life spectacle that never actually took place.[3] Some historians have responded to reenactment with an admixture of bemusement and derision. Greg Dening's irritated dismissal of the practice as the 'present in funny dress' still echoes through the academy; Fred Inglis's more recent characterisation of reenactment as 'dangerous tosh' could become the bookend for Dening's quip.[4] In between, the profession appears to be bent on ignoring reenactment, as if by closing their collective eyes it will go away. It won't.

In the twenty-first century, advances in Computer-Generated Imagery (CGI) and other forms of digital special effects have made it possible to depict people and events of the past with seemingly literal and living detail. CGI versions of historical figures that are grafted onto actors can convey an illusion so real as to be indistinguishable from archival footage. Does this make historical representation more authentic? What does it mean for those who wish to patrol the borderlines between fiction and historical truth? How do we distinguish between a documentary and a fiction clothed in the compelling props of 'reality'? Art and the law provide suggestive pointers to a related ontological problem. One of the unintended outcomes of postmodern appropriation has been that the appropriated artwork can often usurp the original, supplanting it in the mind of the viewer. Similarly at a time when reenactments are increasingly being used in courts of law, there is evidence to suggest that jurors are liable to confuse the accused and the re-enactor. So too with reenactment in other contexts. Those who witnessed a nineteenth-century performance of the trial of an Irish rebel or watched the latest Hollywood version of the life of an Australian bushranger may not only 'believe' what they saw, but they are also likely to substitute crucial aspects of the reenactment for their previously held understandings of the original. It would be a mistake to underestimate the incredulity of the audience but at the same time we must recognise the seductive power of what is before their eyes. It is more than 2000 years since Jesus told Thomas, 'because thou hast seen me, thou hast believed', but the point regardless of medium remains much the same.[5]

The burgeoning popularity of a new 'bastard' genre called 'historical reality television' is also attracting vast viewing audiences as well as waves of disquiet from scholarly historical quarters. New forms of television are being increasingly linked with a long tradition of Western enthusiasm for popular reenactments such as mock battles, heritage scenes and living history museums using meticulously recreated material environments and costumed actors. As the late Raphael Samuel put it, 'Objects must be seen and felt and touched if they are not to remain inanimate ... events should be re-enacted in such a way as to convey the lived experience of the past'.[6]

Arguably, then, reality reenactment in some variant has become the most widely consumed form of popular history. As John Brewer notes in his essay, reenactment is like sexually transmitted disease, there is a lot of it about these days. It also has a fundamental impact on education. A recent survey of history departments in the UK, for example, suggested that the vast majority of students enrolling in undergraduate courses in history had gained their knowledge of history almost exclusively from television and film.[7] Under the rubric of 'experiential learning' or institutions and commercial sites of historical tourism, they dress up, role play and 'experience' the past as part of the syllabus with little or no reflection on the nature of the 'knowledge' they gain. At the Jorvik Centre in York, for example, visitors proceed down flights of darkened stairs to the scene of a former archaeological dig in order to take what might be called a smelly dodgem car ride through a reconstructed Viking village.[8] Among the unconvincingly clothed manikins (à la shop window) are 'authentically' dressed re-enactors, volunteer enthusiasts designed to augment the experience. The re-enactors are passionate about their contribution and, when questioned, expansive about its educational role. Management obviously regards them as asset, featuring them on the cover of a publicity leaflet with the unlikely invitation to become a 'time traveller' (Figure 1.1).[9]

The public presentation of the past then is dominated by reenactment: from 'reality television' costume dramas to weekend medieval tournaments performed by amateur enthusiasts. What lies behind this appetite? What are its implications for the more traditional imperatives of history as an inquiry into the past that seeks explanation and understanding rather than sympathetic and empathetic identification? Is mimetic realism an essential component of reenactment?

When R. G. Collingwood argued famously in defence of reenactment as an essential dimension of good historical practice, he meant that the historian should make a strenuous imaginative leap into the past, trying

Figure 1.1 Jorvik Centre, Publicity leaflet, York, England, 2008

as far as possible to obliterate contemporary knowledge and values: s/he should try to become a medieval monk or a victim of sectarian violence however impossible the aspiration. But he regarded this as a purely intellectual activity. He did not envisage re-enacting the past emotionally or physically – the historical equivalent of archaeological fieldwork – nor could he have imagined the potential of technology to aid the imaginative process.[10] But is this imaginative process enhanced or derailed by our modern capacity to immerse ourselves in a hyperrealistic visual simulation of past environments with the aid of CGI and virtual reality immersion technologies? If computer gamers can replicate the 'magic bullet' that struck Kennedy, does that change the way we regard the findings of the Warren Commission?[11]

Conversely, is the modern form of television reality reenactment compatible with the type of imaginative, rational or romantic interpretations that Collingwood advocated? In practical terms it is also worth exploring if the new forms of virtual reenactment have anything to teach professionals about how to teach, practice or explore the intellectual problems of representing history? If reenactment, whether physical or virtual, can tell us something about context, provide a useful guide to possibilities and help us to appreciate the importance of material culture in shaping experience, can it also help us glimpse the reality of the past?

This book is offered as a contribution to the quest for answers, an opening gambit of sorts. The subtitle, *From Realism to the Affective Turn*, is not necessarily intended to imply a chronology – 'from' is a slippery preposition – although some contributors (and others) would suggest that its chronological implication is apt. As Foucault reminded us in 1971, the genealogy of words and ideas ought to be based on a 'vast accumulation of source material' and 'demands relentless erudition'.[12] The key terms which comprise the title of this volume have different chronologies: 'realism' was coined well over a century ago whereas the ink is scarcely dry on the first use of the phrase 'affective turn'. Space permits neither a voluminous compendium of sources nor uncommon insight, but both terms merit further exposition.

In his seminal *Keywords* Raymond Williams notes the 'almost endless play in the word' 'realism'.[13] As he shows, from as early as the fifteenth century, the word 'real' was used both to denote something actually existing (in contrast to the imaginary) as well as something innately real (most obviously the hand of God). Realism, he suggests, was a term derived from French and first used in the 1850s in Britain where it quickly gained a number of different senses. One of these, the most controversial, is the most relevant to our use of the term here: 'a commitment to describing real events and showing things as they actually exist'.[14] The controversy is useful to pursue. It arose, in part, because realism was both a method and an attitude, the one not always coterminous with the other. More importantly Williams shows the lingering sense that realism might capture that which is beyond the literally real. Here he invokes the Platonic reference in Shelley's *Prometheus Unbound* to 'forms more real than living man'. By the time Shelley put pen to paper, his fellow Romantics, as noted, had flatly rejected literalist simulation, essentially ignoring this use of the term. Nor were Romantics the only ones to be dismissive of this distinction; the curmudgeonly un-Romantic Thomas Carlyle referred insouciantly to Frederick the Great's 'real sins'.[15] The study of reenactment, unlike the practice which is mostly dominated by quirky 'time travellers' and those referred to as authenticity Nazis, slips between the two 'reals'; a desire to learn from the literal recreation of the past and, at the same time, a yearning to experience history somatically and emotionally – to know what it felt like.

A more recent concern catalogued by Williams has direct bearing on the practice of reenactment. This objection concerns the medium by which the real is represented. According to this view it matters not what the medium is – 'language or stone or paint or film' – a superficial representation is a 'falsification making us take the forms of representation as *real*'.[16] With his notion of the hyperreal, Jean Baudrillard in many ways takes up where Williams leaves off. Writing in 1985 Baudrillard argued that in the postmodern world the image is no longer an obviously artificial place marker for the real thing. On the contrary, the simulation – the simulacrum – is more real than the thing itself. The example is not Baudrillard's, but think of the bewildering absurdity when Charlie Chaplin, possessed of arguably the most famous face in the world at the time, entered a Charlie Chaplin look-alike contest and did not even make the final. Simulacra are really real. At the same time the 'really real' is not real at all, or at least not necessarily real, a deliberate ambiguity. For Baudrillard, there was none of the certainty evident in Carlyle's bilious characterisation of a 'wretched Simulacrum' as a 'hungry Imposter, without eyes or heart, practicing for a mess of pottage ... blasphemous swindlery'.[17] Whatever else it might denote, Baudrillard's notion of a hyperreality comprised of simulacra is a place in which reenactment is flourishing.

The noun 'affect' is venerable by antiquity but the fact that it does not appear in Williams' *Keywords* is an indication of how recently it has entered – or perhaps re-entered – scholarly discourse in the humanities. Affective (also not discussed by Williams) is another word with a long etymology and an examination of its use elsewhere in the academy – in cognitive psychology, for example – would open up far too many lines of enquiry than is possible or advisable to pursue here.

The call for an 'affective turn', one of many scholarly turns since the 1980s, was first issued from within media, gender and cultural studies.[18] At a political level, as Lauren Berlant argued in 1997, 'the politics of intimacy', riding on the back of the 'affective turn' has 'usurped the public sphere as a space for social antagonism and struggle, reducing citizenship to personal acts and values, and reframing nationality as a question of feelings and traumas'.[19]

For humanists 'affect' and 'emotion' are, rightly or wrongly, more often than not treated as synonymous. The sense in which 'affective' is used in the title of this book – 'having the quality of influencing the emotions' – is regarded by Oxford's renowned lexicographers as obsolete, a further indication of its recent emergence in academic discourse. The impact of the 'affective turn' on study of history did not occur until early in the new millennium. Affective history as a term was, if not coined, certainly taken up by Chicago scholar Vanessa Agnew, a general editor of this series. Under the banner of the 'affective history' marches the study (if not the practice) of reenactment, sometimes at the forefront, sometimes out of step. Reenactment has a long – if contested – history as an activity, but it too has only recently become the object of scholarly enquiry.

Agnew had participated in the ill-fated (from an ontological viewpoint) reenactment of the first voyage of James Cook's *Endeavour* that was screened on the British national broadcaster in 2001, an experiment described by one of the present editors, Iain McCalman, himself a participant-victim, as 'the little ship of horrors'.[20] For a small group of scholars, the searing experience of *The Ship* sparked an interest in 'extreme history'. Did the 'affective' encounter with pain, suffering (and the potential for death), narrow the distance between then and now?

Among many fruitful trajectories of this initial engagement with the possibilities of reenactment was a typology developed by the *éminence grise* of reenactment studies, Jonathan Lamb. Lamb, another *Ship* survivor and a general editor of this series, divided reenactment into four types: house, pageant, theatre and real. House reenactment, he tells us, 'concentrates on the awkward little things that define a historical moment'. Its interest lies in the 'accumulating frustrations suffered by the re-enactors', much like the reality television shows which have dominated programming since 2004. By contrast the distinguishing feature of 'pageant' reenactments for Lamb is the 'affirmation of a sense of community or nation'. These are not unlike the 'commemoration ceremonies' described by Paul Connerton as 'ritual reenactments' having a 'cardinal importance in shaping communal memory'.[21]

According to Lamb, 'pageant' reenactments range from the sublime to the farcical. The most obvious example of the former is a passion play; history offers numerous examples of the latter. Think of the Duke of Eglinton's tournament in August 1839 when, much to the amusement of the public prints, a number of Tory would-be 'knights' (and others, including the future Napoleon III), spent a fortune to re-enact a medieval joust in rain-soaked Ayrshire.[22] For Lamb 'theatre' reenactments are more intimate than 'pageants' and involve 'a certain amount of extravagance ... in order to lift history above of literal fact, but not quite to the level height of spiritual, national or communal pageantry'.[23] The definition of 'real' reenactments Lamb leaves to a Confederate re-enactor featured in Toby Horwitz's *Confederates in the Attic* who has perfected the art (or artifice) of being a bloated corpse.[24]

The lines between these types are clearly not hard and fast, and they were all forged in the same fire. Lamb's early thinking on the subject almost invariably focused on the 'extreme' through the eyes, variously, of Demodocus, Odysseus, R. G. Collingwood, Jane Austen, W. G. Sebald and J. M. Coetzee's alter ego, Elizabeth Costello. From Austen's *Northanger Abbey*, for example, Lamb interrogates the understanding of history upheld by the novel's heroine, Catherine Morland. For Morland, the attraction of the past is its capacity to open up the pain of its subjects to sympathetic understanding. The 'proper adjustment of pain to sympathy', Lamb opines, 'seems to provide the affective basis of historical reenactment'.[25]

Soon, however, a second wave of interest brought other scholars and other subjects to the conference table – pleasure, art, science, poetry, music, film, politics, the prosaic, the everyday. Consciously or unconsciously then 'extreme history' became 'affective history'. Here we have intentionally left the definition of reenactment loose and the range of genre generous – indeed it is central to our objectives to help explore its definition and its boundaries. Readers who find our use of the term too capacious and our perimeters too porous are invited to offer tighter or narrower definitions in future volumes.

All the essays in the volume engage, in one way or another, with the quest for a definition and an appropriate delimitation of the subject as well as the philosophical and practical questions revolving around the vexed relationship between realism and affect. Without heading into an epistemological cul-de-sac, it is worth recapitulating our critical questions. Is reenactment the key to knowing what it was like to live in the past? Would the re-enactor's holy grail of realism or authenticity narrow the gap between past and present so that we might touch it? Exploring these questions is aided by attention to genre and, to that end, many of the essays here explore the place of reenactment in pursuits within and beyond the academy – history, literature, music, theatre, dance, the law, film, television, public commemoration and historical tourism. A couple of examples will suffice here. Sites of historical tourism – public and commercial – (discussed here by several authors) are intrinsically concerned with creating an affective relationship with the past because of the very fact that this past is objectified for the visitors. Some have also dabbled with the use of sensory experience to augment this. But how do we know what the past felt like; how it smelled, tasted, sounded? As Norbert Elias has shown in relation to social etiquette, including smell, the 'threshold of repugnance' is socially constructed and varies greatly between cultures and over time.[26] John Brewer also makes this point in relation to neo-realist cinema that the assumption that somatic experience and feeling are timeless is deeply flawed.

The uneasy relationship between realism, authenticity and affect is further evident in cases where the boundaries of reenactment are pushed towards improvisation. By allocating visitors unscripted parts in otherwise professionally staged reenactments, tourist sites have been at the forefront of this strategy. How much historical understanding is advanced by this process is debatable, but turning tourists into 'spect-actors' (to borrow a term from Augusto Boal[27]), with all its Brechtian connotations (drama-as-a-medium), has perhaps got more to do with entertainment than pedagogy. On the other hand, Katie Kitamura shows in her chapter that mixing re-enactors and the historical actors themselves can reveal an instructive paradox. Exploring Jeremy Deller's staged reenactment of an infamous clash between picketing miners and the police that had taken place 17 years earlier in Mrs Thatcher's Britain, Kitamura shows that the re-enactors in this extraordinary piece of performance art were intent on recreating the event accurately down to the last blow, while the veterans of the clash itself were only too happy to rewrite history. In his chapter Stephen Gapps, however, shows that unschooled re-enactors can sometimes lose their heads and attempt to change history. In a scene reminiscent of the well-known episode of Matt Stone and Trey Parker's *South Park* in which Cartman foments a rebellion among Confederate re-enactors who go on to 'win' the Civil War, Gapps shows how, as a professional re-enactor, he almost lost control of those re-enacting the part of the Irish rebels at Castle Hill in New South Wales in 1804.

Taking up much the same issue in his examination of Rossellini's cinema, John Brewer insists that students of reenactment need to find a way of dealing with contingency and chance. In the interests of conceptualising a poetics of reenactment Brewer suggests his own typology that contrasts with Jonathan Lamb's (and Connerton's) in important respects. In the light of the way that Rossellini's films explore what he calls the dialectic of the bigness of history and smallness of everyday life, Brewer goes on to call for an appreciation of the difference between historical and poetic truth. A similar issue is canvassed by Paul Pickering in his discussion of the challenge thrown down by noted film historian Robert Rosenstone.

Rosenstone argues that, unlike historians who cannot break the bounds of linear narrative, filmmakers are able to create a past that 'is at once serious, complex, challenging, and "true" in its ability to render meanings rather than the literal reality of past events'.[28] Pickering examines a number of films and television programmes devoted to an iconic nineteenth-century Australian rebel and finds that none of them fulfil Rosenstone's promise, offering instead a range of distortions (intentional and otherwise) driven mostly by the desire to exculpate or profit. Nevertheless this investigation underscores the enormous power of re-enacting the past on the screen (large and small). Historical films, regardless of their concern with authenticity, are surely the quintessential simulacra of Baudrillard's imagination. The invitation to spectators to believe their eyes is also a conclusion drawn by Iain McCalman in his chapter on de Loutherbourg.

Another point at which even poetic truth shades off into fiction is also the subject of consideration in the volume. In their chapters both Anja Schwarz and Paul Pickering discuss the controversy that engulfed the Australian novelist Kate Grenville when she propounded an 'experiential' theory of writing. Grenville was scorned by the academy both for her shortcomings as a researcher and for her methodological innovation, which is actually reenactment by another name. Two issues emerge from this debate: reenactment as a form of redemption and reenactment as a methodology (to which we will return).

This question of redemption emerges strongly in Schwarz's essay as well as in James Walvin's. Examining Grenville's controversial novel, *The Secret River*, alongside an Australian television series titled *Outback House*, screened on the Australian public broadcaster in 2005, Schwarz notes the fixation with the moment of colonisation and an underlying sense of unease about belonging. Both the show and the novel, she notes, are concerned not with understanding the past but with rectifying past wrongs in the present. In this respect Schwarz highlights Grenville's use of knitting as a metaphor, as if history can be unravelled back to a 'mistake' – in this case colonial dispossession – and re-knitted. In *Outback House* reenactment is the vehicle for this anachronistic redemptive narrative. Again, it is worth noting that film and television have proved to be highly popular sites for lingering over reenactment in the presence of a mass audience. Following *South Park*, another popular animated comedy, *Family Guy*, featured a Civil War reenactment where the outcome was contested. The highly successful situation comedy

Everybody Loves Raymond used the Civil War as a crucible not to rewrite the past but to explore the affective capacity of reenactment in the present:

RAYMOND: Either I'm on your side in this thing or I'm not going to go.
FATHER: You can't desert now.
Raymond: You can't desert Dad, there's no war here. It's just dressing up in costumes like it's Halloween with your friend Harry-slash-Jebb.
FATHER: You shut up now. This means something to us. There's camaraderie here. There's spirit. You don't understand anything about this. You have no idea what it feels like to stand on a battlefield pretending to fight for something you believe in.[29]

In each case the object was not trauma but an almost insidious sense of discomfort disguised by humour.

At a more general level Jonathan Walker in his chapter points to the potential of reenactment to act as a metaphor. This comes close to another intriguing paradox: actualising the past as signifier of something else; resolution or understanding via a reenactment of a reenactment. Stephen Gapps also makes a case for the cathartic value of reenactment and he goes on to suggest that as a participatory activity, it is at the cutting edge of what he calls democratic history. Clearly re-enacting for the benefit of the present is beneficial to both audience and re-enactor but, at the same time, popularity is an odd index of historical accuracy. Regarding history as 'unfinished business', Gapps tells us, is useful at getting to the heart of unresolved issues; it might also be taken as an invitation to make it up.

On this question of the tension between creativity and authenticity James Walvin suggests in his essay that one of the central functions of many sites of historical tourism is to assuage the collective guilt of the history's victors (in this case the beneficiaries of the slave trade). Walvin shows, for example, that in tourist sites devoted to plantation history reenactment and role-playing of various kinds are standard fare. These pageants are almost invariably popular with visitors except when they are used to disrupt the dominant narrative. Telling both sides of the story, Walvin notes, has proven to be discomforting to descendents of the victims and perpetrators alike. This conclusion suggests that the generic consumer of public history is less concerned with a warts-and-all understanding of what happened than with closure even if this means accepting a sanitised version of the past. Affect is not universally desirable; joy is welcome, revulsion less so. It is surely no accident that the so-called history wars that have raged in many countries are almost invariably about what citizens feel about their past – pride or shame.[30] Given that sites of historical tourism, as with historical films and television shows, are firmly lodged in the education system, the implications of this are considerable.

A third important strand in many of the essays is that they focus on the history of reenactment. This is a neglected area. For all that, new developments in aesthetics and technology have produced a remarkable shift in the modes of presenting history; the practice of reenactment itself has, as we have noted, a long history. Historicising reenactment undoubtedly helps us to understand it. For example, both in the postmodern world and in the past, reenactment has been crucial to the creation and contestation of public memory. Part of the task of historians is to explore its various manifestations and to track its development over time, which, in itself, raises a number of contentious issues. The final two essays written by Simon During and Iain McCalman respectively, for example, traverse some of the same historical ground but offer different conclusions about the beginning of what we might call modern Western reenactment. McCalman suggests that its inception is to be found in the experimental spectacles created by Philippe de Loutherbourg in late Georgian London; for During these were merely part of series of proto-reenactments that presaged the formal commencement of the practice in the aftermath of the American Civil War. By asserting that modern reenactment began in London in the 1780s, McCalman directly challenges the popular US view that the fiftieth anniversary reenactment of the battle of Gettysburg heralds the beginning of the genre. His suggestion that de Loutherbourg presented the first disaster movie also contradicts numerous later claims to originality.

The relationship of reenactment to the historian's craft is as much a practical question of method as it is a philosophical conundrum. This is another issue raised in several chapters, including those which muse over R. G. Collingwood's ideas. By exploring Collingwood's early writings about music, Kate Bowan draws attention to a fundamental contradiction in his thinking about the practice of reenactment. As noted, in relation to history, Collingwood ultimately envisaged reenactment as an entirely cerebral activity, but here we meet Collingwood the 'man of action', or as Bowan puts it, the man who believed that the reenactment of the past as lived experience was important, who maintained that a student of music could not imagine a Beethoven symphony without having heard it performed. The issue of reenactment as an embodied activity is also examined by Paul Pickering. By examining the extensive use of reenactment in the criminal justice system – both inside and outside the courtroom – Pickering notes that a similar range of issues are exercising the minds of scholars. Although re-enacting is increasingly seen as problematic when brought into the court as evidence, its potential is embraced willingly by criminal investigators without any of the angst displayed by historians. As an avid reader of detective fiction, Collingwood knew this well; by restricting reenactment to his imagination, he forgot it. The debate about the place of reenactment within the law is clearly an arena for future exploration.

So too with other disciplines.[31] In her essay, Katie Kitamura draws on cultural theory – in particular Nicolas Bourriaud's notion of 'relational aesthetics' – to consider performance art as reenactment. Similarly, Ruth Phillips and Tracey Nicks highlight the potential for the study of reenactment through dance. In her chapter on music, Kate Bowan explores formative attempts by a pioneer of the Early Music Revival to recreate early Western music while festooned in Elizabethan pantaloons before appreciative audiences of the Bloomsbury cognoscenti. She then goes on to link this to the fierce debate among classical musicians about the possibility of authenticity within historical music performance. This debate has raged for decades and, as Bowan shows, it offers important insight to students of reenactment on the tension between creativity and authenticity and the limits and possibilities of the latter. A further contribution to the discussion of reenactment as a methodology is Iain McCalman's notion of 'reflexive reenactment'. Accepting the fact that re-enacting can never fully capture what it might have felt like to be there, he has argued that we should make a virtue of that shortcoming. The very element of unpredictability, he suggests, can become a source of creative exchange with the past, provided it is frankly acknowledged.[32] Taken together the chapters in this volume suggest that taking reenactment seriously as a methodology is worth the risk and that its potential is best explored through an interdisciplinary lens.

Finally the book explores the extraordinary potential of reenactment as a form of extra-textual communication. It is clear from the essays by Phillips and Nicks, Walker and Bowan that reenactments have a powerful and immediate impact on a visceral register and can reveal the past in ways that words cannot.[33] Phillips and Nicks demonstrate how Esther Deer, an acclaimed Hodenosaunee (Iroquois) entertainer who performed in Europe, Africa and North America under the name of Princess White Deer, used reenactment to destabilize potentially paralysing stereotypes of 'Indianness' current during the early twentieth century. Through dance and costume, Deer conveyed a form of intentional hybridity at a time when Indian identities faced the challenge of cultural evolutionism and the belief that distinctive Native cultures could not survive. This was reenactment as a form of protest without words. Similarly the images Walker discusses in his chapter highlight both the difference between somatic experience and text and the potential of the non-verbal forms of reenactment.

At one level using reenactment might be seen to emerge from our enquiry as little more than the historian's equivalent to archaeological or anthropological fieldwork, an attempt to take up R. H. Tawney's well-known injunction to historians to lay aside their books in favour of a stout pair of boots.[34] The value of this conclusion should not be gainsaid. The noted biographer Richard Holmes has given us a vivid example of the value of reenactment, although he does not use the word, preferring what he calls 'foot-stepping' to describe his method. In his essay on Mary Wollstonecraft, Holmes writes of his dismay at reading a line in one of her letters in which she claims to have seen the king being taken to his execution in 1793 from her apartment window.[35] Having done his homework Holmes knew that the hapless Louis had travelled down a different street, one running parallel to Wollstonecraft's, on his way to the guillotine. The king's route had been close enough to form the basis of a good tale for her correspondent but was clearly false. This revelation profoundly shook Holmes' faith in his subject: if she lied about this, what else had she lied about? Was this proof positive that her letters were tainted? In Paris, tracing her footsteps, Holmes went to her apartments, and looked from the window that had formed the basis of her fallacious account. From here, however, it became clear that she *could* indeed have seen over the rooftops below to the king in doleful progress. Wollstonecraft was vindicated. Holmes' attitude to his subject was profoundly shaped by reenactment, a ringing endorsement of its value as a historical method.

At the same time, exploring the motives and intentions of re-enactors can provide a useful window into shifting cultural practices and contested values. If the gap between past and present is not narrowed by the mimetic realism of embodied reenactment, the study of reenactments at fixed points on the continuum between then and now helps us to map change over time and to understand the lived experience of those doing the re-enacting. Moreover the task of historians is not only to seek to understand the past, but also to engage with the ways that it speaks to the present. In one form or another reenactment is here to stay. It demands our ongoing attention. In many ways then this volume is intended to set an agenda for taking reenactment seriously; to open a page, to raise a range of issues, and to invite others to join in the discussion.

Notes

1. See http://www.nares.org.uk/main_pages/Re-enactment.htm.
2. See T. Horwitz (1998) *Confederates in the Attic: Dispatches Form the Unfinished Civil* (New York: Vintage).
3. R. Barthes ([1968] 1982) 'The Reality Effect' in *French Literary Theory Today: A Reader*, ed. T. Todorov (Cambridge: Cambridge University Press) 11–17; U. Eco (1986) *Travels in Hyperreality: Essays* (San Diego: Harcourt Brace Jovanovich); J. Baudrillard (1995) *Simulacra and Simulation* (Ann Arbor: University of Michigan Press).
4. G. Dening (1992) *Mr Bligh's Bad Language* (Cambridge: Cambridge University Press), pp. 4–5; F. Inglis (2008) 'Re-enacting a Life: On Writing the Biography of R. G. Collingwood', paper presented to Research School of Humanities, The Australian National University, 9 May 2008. See also M. H. Nielsen (1981) 'Re-Enactment and Reconstruction in Collingwood's Philosophy of History', *History and Theory*, 20:1, p. 3.
5. John 20: 30, *Bible* (King James Edition).
6. R. Samuel (1995) *Theatres of Memory* (London: Verso), pp. 175–6.
7. See R. Pearce (2003) 'University History', *History Today*, 53:8, p. 54. See also G. Wong (2008) 'Educators Explore Second Life Online', http:www.cnn.com/2006/TECH/11/13/second.life.university; T. Hunt (2006) 'Reality, Identity and Empathy: The Changing Face of Social History Television', *Journal of Social History*, 39:3, pp. 843–58.
8. The characterization is Kate Bowan's.
9. Chapter 1 of B. Gibbon (2005) *The Colony: The Book from the Popular SBS Living History Series* (Sydney: Random House) is entitled 'Time Travellers'.
10. R. G. Collingwood ([1946] 1971) *The Idea of History* (Oxford: Oxford University Press).
11. Designed by Scotland-based Traffic Games and released in 2004, 'JFK Reloaded' is described as a '3-D, First-Person Perspective, "Assassination Simulator"'. The player sits in the infamous Texas book repository with a gun and from the same vantage point as Lee Harvey Oswald attempts to replicate the magic bullet. Initially a cash prize was offered to anyone who could do so. According to the game designers it was intended to silence conspiracy theorists. See http://www.cool.com.au/computers-technology/personal-computers/jfk-reloaded-revisited-20060414256/. We are grateful to Lucinda Mathieson for this reference. See also J. de Groot (2006) 'Empathy and Enfranchisement: Popular Histories', *Rethinking History*, 10:3, pp. 391–413; B. Rejack (2007) 'Toward a Virtual Re-enactment of History: Video Games and the Recreation of the Past', *Rethinking History*, 11:3, pp. 411–25.

12. M. Foucault (1977) 'Nietzsche, Genealogy, History' in *Language, Counter-Memory, Practice: Selected Essays and Interviews*, ed. D. Bouchard (Ithaca: Cornell University Press), p. 140.
13. R. Williams ([1976] 1983) *Keywords: A Vocabulary of Culture and Society*, ed., rev. and expanded (London: Fontana Paperbacks, Flamingo), p. 258.
14. Williams, *Keywords*, p. 259.
15. T. Carlyle (1872–3) *History of Friedrich II of Prussia, called Frederick the Great*, vol. 2 (London: Chapman and Hall,), p. 558.
16. Williams, *Keywords*, p. 261.
17. T. Carlyle (1868) *On Heroes, Hero-Worship and the Heroic in History* (London: Chapman and Hall), p. 60.
18. V. Agnew (2007) 'History's Affective Turn: Historical Re-enactment and Its Work in the Present', *Rethinking History* 11:3, p. 310. See also A. Cook (2004) 'The Use and Abuse of Historical Re-enactment: Thoughts on Recent Trends in Public History', *Criticism*, 46:3, pp. 487–96.
19. A. Koivunen (2001) 'Preface: The Affective Turn?' in A. Koivunen and S. Paasonen (eds), *Conference Proceedings for Affective Encounters: Rethinking Embodiment in Feminist Media Studies*, University of Turku, Series A, no. 49, p. 1. See http://www.utu.fi/hum/mediatutkimus/affective/proceedings.pdf.
20. See I. McCalman (2004) 'The Little Ship of Horrors: Re-Enacting Extreme History', *Criticism*, 46:3, pp. 477–86. See also S. Baker (2002) *The Ship: Retracing Cook's Endeavour Voyage* (London: BBC Worldwide); A. Cook (2004) 'Sailing on The Ship: Re-enactment and the Quest for Popular History', *History Workshop Journal*, 57, pp. 247–55.
21. P. Connerton (1989) *How Societies Remember* (Cambridge: Cambridge University Press), p. 61.
22. See I. Anstruther (1963) *The Knight and the Umbrella: An Account of the Eglinton Tournament, 1839* (London: G. Bles). We are grateful to Alex Tyrrell for this reference. The Eglinton tournament was itself re-enacted in 1989.
23. Lamb has outlined this typology at numerous conferences and seminars since 2001. It appears in published form in his most recent book. See J. Lamb (2009) *The Evolution of Sympathy in the Long Eighteenth Century* (London: Pickering and Chatto), p. 137. Paul Connerton also offers a useful typology: 'calendrical, verbal and gestural'. See Connerton, *How Societies Remember*, p. 65.
24. Lamb, *The Evolution of Sympathy*, p. 139; see T. Horwitz (1998) *Confederates in the Attic*, p. 8. See also D. Hall (1994) 'Civil War Reenactors and the Postmodern Sense of history', *Journal of American Culture*, 17:3, pp. 7–10.
25. Lamb, *The Evolution of Sympathy*, pp. 133–4.
26. See N. Elias (2000) *The Civilizing Process: Sociogenetic and Psychogenetic Investigations* (Oxford: Oxford University Press). We are grateful to Michael Davis for this reference.
27. A. Boal (2002) *Games for Actors and Non-Actors* (New York: Routledge). See also S, Magelssen (2006) 'Making History in the Second Person: Post-Touristic Considerations for Living Historical Interpretation', *Theatre Journal*, 58, pp. 291–312; R. Blackson (2007) 'Once More ... With Feeling: Re-enactment and Contemporary Art and Culture', *Art Journal*, 61:1, pp. 36–7.
28. R. Rosenstone (1995) *Visions of the Past: The Challenge of Film to our Modern Idea of History* (Cambridge, MA: Harvard University Press).

29. *Everybody Loves Raymond* (1998) Season Two, *Civil War* 14:55. We are grateful to Suzanne Pickering for bringing this to our attention.
30. See for example S. Macintyre and A. Clark (2003) *The History Wars* (Melbourne: Melbourne University Press). See also S. Gapps (2003) 'Authenticity Matters: Historical Re-enactment and Australian Attitudes to the Past', *Australian Cultural History* 23, pp. 105–16; A. Schwarz (2007) 'Not This Year! Reenacting Contested Pasts Aboard *The Ship*', *Rethinking History*, 11:3, pp. 427–46.
31. The same is true of other disciplines. Drawing on the work of Michael Taussig some social anthropologists employ what they call an 'intuitive' approach in the face of fraught fieldwork conditions. Monique Skidmore, for example, has argued that the way to study the lives of those who live in an oppressive regime is to place yourself in danger alongside your subject and write about how you feel. To understand the 'affective dimensions of dictatorship', she writes of working in the field in Myanmar, 'I intuit the experiences of the Burmese people I have come to know well'. Her terror, she claims, allows her to experience theirs intuitively without having to name them. See M. Skidmore (2003) 'Darker than Midnight: Fear, Vulnerability, and Terror making in Urban Burma (Myanmar)', *American Ethnologist*, 30:1, pp. 5, 7.
32. See I. McCalman, 'Past Not Just the Present in Silly Clothes', *Sydney Morning Herald* (17 September 2007).
33. See also Connerton, *How Societies Remember*, pp. 72–104.
34. Cited in E. Kerridge (1951) 'Ridge and Furrow and Agrarian History', *Economic History Review*, New Series, 4:1, p. 14 and 14n.
35. R. Holmes (1996) *Footsteps: Adventures of a Romantic Biographer* (New York: Vintage Books).

2

'... Just as It would have been in 1861': Stuttering Colonial Beginnings in ABC's *Outback House*

Anja Schwarz

What might have been is an abstraction
Remaining a perpetual possibility
Only in a world of speculation.
What might have been and what has been
Point to one end, which is always present.
Footfalls echo in the memory
Down the passage which we did not take
Towards the door we never opened
Into the rose-garden.

—T.S. Eliot 'Burnt Norton'. *Four Quartets*

In July 2005, Australian novelist Kate Grenville was invited by Radio National's Books and Writing programme to talk about *The Secret River*, her latest novel loosely based on the life of her ancestor Solomon Wiseman, who had been transported to the penal colony of New South Wales in 1817 and later settled on the Hawkesbury River.[1] The show's host, Ramona Koval, congratulated Grenville on the poetic tone and language of her narrative and praised the text as 'a wonderful and disturbing novel, full of detail about life and work in the colony ... and daring descriptions of the land and the strangeness of the encounters between black and white people'.[2] Grenville, in turn, read a passage from the book and reflected comprehensively on what had motivated her to write it. Then came Koval's final question and Grenville's reply which would irritate and preoccupy the country's historians for months to come: 'So, where would you slot your book', Koval had enquired, 'if you were laying out books on the history wars? Whereabouts would you slot yours?' Australia's 'history wars', heated debates among historians and public intellectuals over the nature of the country's colonial legacy,[3] had over the previous decade caused a deep rift between the so-called black armband historians on the political left and conservative scholars accused of wearing a 'white blindfold'[4] on the right. Grenville, it seems, sought to remain out of the historiographical crossfire when she replied:

> Mine would be up on a ladder, looking down on the history wars ... [A] novelist can stand up on a stepladder and look down at this, outside the fray, and say there is another way to understand it ... actually experiencing what it was like, the choices that those people had ... The historians are doing their thing, but let me as a novelist come to it in a different way, which is the way of empathising and imaginative understanding of those difficult events. Basically to think, well, what would I have done in that situation, and what sort of a person would that make me?[5]

Grenville's statement consists of two interrelated lines of argument which brought to the defences a number of historians, weary, according to Mark McKenna, of the 'rise of the novelist as historian, of fiction as history'.[6] Firstly her claim to a different, possibly superior, perspective on the past, they felt, had illegitimately crossed the line between imaginative fiction and evidence-bound history. Inga Clendinnen, Grenville's most prominent critic, objected accordingly that the author had trespassed into alien 'territory' and tried to 'bump historians off the track'.[7] Grenville has since explained that – when looking for an adequate metaphor to convey her understanding of the novelist's role in the rush of the radio interview – she did not intend the step-ladder imagery 'to imply superiority'. She had rather sought to represent her own approach to the past as distinctly literary, 'perched up high on a removed vantage point where I could watch [the history wars] but not be involved'.[8]

The second part of Grenville's reply to Koval further explored this notion of a 'different' access to the past. It argued that the novelist's and, by implication, the readers' experience and subsequent empathy with historical actors, opened up another 'track back into the past'[9] different from that of historical scholarship which allowed them to imagine how they would have acted under the same circumstances. Clendinnen, again, was not happy: 'Grenville would not have been Grenville in "that situation"', she pointed out, 'We cannot post ourselves back in time'.[10] In his *Sense and Nonsense in Australian History*, published shortly after Grenville's radio interview, John Hirst had argued similarly that 'Kate Grenville cannot imagine how she would have behaved on the Hawkesbury frontier ... The pioneer settlers are not ourselves'.[11]

I have begun this essay with the recent debate about *The Secret River* because Grenville's statements, as well as the angry criticism flung at her, touch upon epistemological concerns very similar to those that preoccupy historical reenactment. These performances of the past share with Grenville habitual assertions of a different and possibly privileged access to the events they revisit, and they rely equally on experience to verify these claims to historical accuracy. The following statements made by two participants of *Outback House* (Australian Broadcasting Corporation, 2005), the television programme that constitutes the focal point of my analysis, reveal a striking similarity both to Grenville's truth claims and to the validating function she accords to experience: '[We are] very rare people who have actually lived in the past, which no-one can really do as we have done', states Bernie, cast in the role of the station's shepherd, in a retrospective interview.[12] And Russell, one of *Outback House*'s station hands, adds: 'What you've got to think about is that we lived it. We actually lived 1861. It was three dimensional for us'.[13]

Outback House belongs to the 'bastard genre'[14] of historical docudrama. A generic hybrid, it mixes traditions of re-enacting historical events in period dress with documentary television and Big Brother's 'fly-on the wall' format that portrays participants subjected to a group endurance test.[15] The ABC promoted the series as an 'incredible experience of living history' that documented how a group of Australians travelled 'back in time to spend three months running a remote nineteenth century sheep station exactly as our forbearers did in 1861'.[16] As evident from Bernie's and Russell's comments, the programme might productively be read together with Grenville's authorial self-disclosure. Their similarities go far beyond a mutual privileging of corporeal experience over historical scholarship: *Outback House* shares with Kate Grenville a specific motivation for engaging with Australia's colonial history and articulate worries of non-Indigenous Australians about their right to belong. This essay is not the first text to comment on Australian reenactments as primarily preoccupied with imperial history,[17] and astute analyses of *Outback House*'s articulation of contemporary settler colonial narratives have recently been offered by Michelle Arrow and Catriona Elder.[18] My analysis will contribute to existing scholarship by exploring the show through the lens afforded by Grenville's statements.[19]

After scrutinising Grenville's notion of experience and investigating the epistemological functions this category fulfils for the novel and for reenactment, my analysis will focus on a peculiar fixation with the moment of colonial beginning that characterises *Outback House* and which also informs Grenville's writing. The essay will examine this phenomenon by paying attention to its particular temporal framing. Like Grenville, when asking what she *would have done* at the Hawksbury frontier of the early nineteenth century, television reenactments make excessive use of the 'past conditional' (would+have+past participle). Exploring the particular historiography facilitated by this temporal mode, I consider its role in articulating a renewed sense of belonging for non-Indigenous Australians. Such narratives in the past conditional, I argue, answer a particular desire for spatial and temporal anchoring in confirming a link between non-Indigenous Australians and *their* continent. In rendering as authentic a non-Indigenous twenty-first-century fantasy of the colonial past, they understand colonial history as always already settled.

2.1 'What it must have been like'

In a number of interviews, Grenville has expanded on the importance of corporeal experience for her own writing as well as for her book's future readers. Seeking to convey the 'actual experience' of the past, she had aimed to make it a text 'you could taste and smell and touch'.[20] Grenville characterises herself as 'a great believer in the experiential theory of writing'[21] and detailed accordingly how she had gained first-hand insights into the past: as often as possible the author had put herself in the position of her historical characters, trying 'to feel what they felt'.[22] From a rag dipped into animal fat, for instance, she made a so-called slush lamp. 'When I turned the lights off and lit it', Grenville relates, 'I suddenly understood what it must have been like in a bark hut at night – the dimness, the darkness, the smell and the smoke'.[23] Sensory experiences such as this one, the author believes, taught her 'more about life in a bark hut on the Hawkesbury in 1817 than all the books in the world could have told me'.[24]

A similarly fundamental dependence on the 'evidence of experience'[25] informs the recent boom of living-history formats on television. In *Outback House*, experience is central to the narrative from the moment the show's participants are imagined to enter the life of their historical personas. Constructing the volunteers as those 'very rare people who have actually lived in the past', it thus puts into practice Grenville's rationale that 'book research can only take you so far'.[26] Whereas historians are named in the closing credits of each episode and lend authority to the historical panorama painted by *Outback House*, the programme's claim to knowledge about the past depends not so much on their expert input but on the re-enactors' bodily immersion in a 'three dimensional' environment of simulated pastness.

In the show's first episode, travelling back in time is configured spatially.[27] The volunteers' access to this environment is barred by a number of thresholds they have to cross before 'arriving' in 1861, the year their historical performance is imagined to take place. The first episode follows the Allcorns, the programme's nineteenth-century squatter family, as they leave their twenty-first-century Sydney home and are brought to the liminal site of a training camp in rural New South Wales. Together with the show's cook, overseer, shepherd and station hands, the family acquires basic farming skills and knowledge of Victorian etiquette. Finally, in the deep of night, they enter a dressing tent that sees them transformed into their historical nineteenth-century characters. The final threshold is traversed when the volunteers shut behind them the gates to the property of Oxley Downs, the show's nineteenth-century sheep station, and, by implication, relinquish all ties to the present.

Once 'arrived' in the past, the sensory experience of the simulacra of Oxley Downs seemingly enables *Outback House*'s participants to bear witness to life in 1861. Their role therefore corresponds roughly to what Freddie Rokem has observed for actors in theatrical performances of history in general: as 'hyper historians' their re-living of the past guarantees and validates the historical authenticity of the events portrayed.[28] And similar to Marita Sturken's observations about the truth value accorded to survivor testimony, we might conceive of re-enactors as 'tell[ing] us the real story, one wrought from experience'.[29] Within the *Outback House*'s narrative the participants, who reside in the simulated environment of 1861, are consequently granted the authority to make truthful statements about life on a nineteenth-century outback sheep station.

Period clothing plays a central role in creating these stories 'wrought from experience' and, as with other reenactment programmes restaging the nineteenth century, the body of the female re-enactor and its subjection to a Victorian dressing routine is central in constructing this narrative.[30] 'Eighteen sixty-one', the show's authoritative voiceover informs us while we see images of female volunteers getting dressed, 'was the age of the tight corset, the crutch-less pantaloon and the looped crinoline. Our wardrobe department has recreated everything down to the finest period detail'.[31] Female garments of the Victorian age are rendered here as regulatory and disciplining contrivances that constrict twenty-first-century bodies. Transforming looks and body posture, they take on the role of time-travel devices. Julie Allcorn comments accordingly on the process of changing into her historical attire:

> I've got my stockings on ... the corset – and this is the beginning. [Putting on an underskirt] Look at this lacework! Unbelievable. [After several layers of skirts] I'm getting bigger, I'm getting bigger here. [Now fully dressed in her Victorian outfit and looking at herself in the mirror] Wow. You feel like a princess. That's beautiful. Absolutely stunning.[32]

Taking a lead from Foucault, one might almost describe Julie's body as 'totally imprinted by history'[33] and produced by the disciplining regime of Victorian undergarments. But whereas Foucault would have understood individual identities as resulting from such inscriptions, *Outback House*'s re-enactors are thought to remain their twenty-first-century selves, 'using' their bodies' subjection to 1861 conditions so as to bear witness to the past. Paul Allcorn conveys this understanding of his transformation when likening his experience of getting dressed to notions of temporal and/or social displacement: 'I feel like I'm going back to Jurassic Park or like I am going to prison for three months', he comments while changing into his outfit. '[It] must feel a little bit like this, the same I would say'.[34]

2.2 'You want to go back 200 years'

Although a critical exploration of the category of experience is vital in explaining the manner in which reenactments construct their claims to historical accuracy, this alone cannot account for their recent popularity nor does such an analysis help understand which perspective on the past is given currency by the recent boom in reenactment television.

Five thousand Australians applied to participate in *Outback House*.[35] Additionally a vast number of viewers engaged with its themes and issues on the show's website where they passionately discussed each of the weekly episodes. Most prominent among these comments were remarks on the programme's success or failure in accurately reproducing the nineteenth century. A significant segment of the audience, it seems, was genuinely disappointed. 'Why is it so hard for people', one of them queried, 'to "think" themselves back to the 1860s and to try to react accordingly?'[36] Their complaints covered a broad spectrum: from ahistorical building designs and equipment to the inappropriateness of the use of first names, greetings or, more generally, a perceived lack of etiquette. Each episode saw a new list of 'mistakes' being drawn up by critical members of the audience, trying to outdo each other, it seems, in their detailed knowledge of the period.[37] These discussions notwithstanding, most viewers appeared to enjoy *Outback House*, sometimes even expressing a sentiment of nostalgia when comparing contemporary Australian society to the nineteenth-century past performed in the programme: 'I suddenly realized that I had lost something living in the 21st century urban Australia, and that is a sense of "community"', wrote 'ASHNIC' on 7 August 2005. 'Despite our modern technology and easy lifestyle "community" is something that seems to have been lost somewhere between 1861 and 2005'.

Comments such as these help understand the current popularity of historical reenactment as part of a broader change in the relationship to the colonial past; a shift that Ian Baucom and Paul Gilroy have mapped for Britain, the former colonial centre, but which is equally at play in ex-settler colonies such as New Zealand and Australia.[38] In fact, memory, as Andreas Huyssen has observed with regards to the universal dimensions of this shift, has become a key concern for societies around the globe. Huyssen argues that this trend makes itself felt not only in increasing efforts to preserve the past – 'Total recall seems to be the goal'[39] – but can also be detected in the growing popularity of alternative versions of historical events, such as those fashioned by reenactments: '"Original remakes" are in, and as cultural theorists and critics we are obsessed with re-presentation, repetition, replication, and the culture of the copy, with or without the original'.[40] Huyssen suggests that we understand the recent significance accorded to commemoration and remembrance as a reaction to processes of deterritorialisation that result from progressive globalisation. The current prominence of memory discourses, he argues, expresses a desire for 'spatial and temporal anchoring in a world of increasing flux in ever denser networks of compressed time and space'.[41]

Against this backdrop of a general spatio-temporal uprootedness, Vanessa Agnew proposes reading the reenactment boom as participating in history's 'affective turn' characterised by a 'concern with personal experience, social relations and everyday life' and preoccupied with establishing 'conjectural and provisional interpretations of the past'.[42] As most contributors to *Outback House*'s online discussion forum understood quite clearly, these provisional, affective interpretations are less concerned with historical accuracy than with the viewers' and re-enactors' present, no matter how often the show promised to deliver 'a fascinating insight into the real-life struggles of the early settlers'.[43] 'Ispy' remarked accordingly that 'This is not a factual or historical documentary about squatters in the 1860s. We are watching modern day people to see how they cope'. And 'ASHNIC', in another comment, added that the interweaving of past and present was precisely what 'makes this programme so interesting'. Rather than engaging with the dubious claims to historical accuracy made by the various *House* formats, Agnew therefore comprehends their relevance in terms of the insights these programmes allow into the nature of contemporary historical discourse and into the presentist agendas that shape reenactments' respective commemorative contexts.[44]

What conjectures, then, does *Outback House* establish between Australia's past and present and whose desire for spatial and temporal anchoring does the popular programme answer? Here, Kate Grenville's reflections on her motivation for corporeal encounters with the nineteenth-century life world of her protagonist again afford a valuable perspective on reenactments. In interviews, along with a more detailed discussion of the project published as *Searching for Secret River*, the author expressed unease about the way in which her own 'sense of "belonging" ... as a fifth-generation Australian' had increasingly become bound up with the country's colonial past. For more than a decade, Grenville explains, her sense of place had grown 'more and more hollow'[45] as pressing political questions concerning the relationship of Indigenous and non-Indigenous Australians were not being answered. She understands this difficult relationship to be rooted in the country's colonial legacy and therefore believes the past to hold the key for a possible future reconciliation: 'in a situation like that, the only thing you can do is go back to the point where it went wrong'.[46] Referring to colonial conflict and interracial violence in the colony's early years, she expands: 'You want to go back 200 years and say to the settlers, "Look, this is how the Aborigines are", and to the Aborigines, "Look, this is why the settlers are behaving the way they are. Let's understand this. There's no need for all this brutality"'.[47]

Understandably historians, and prominent among them Hirst, McKenna and Clendinnen, were scandalised. '[I]f Aborigines had earlier understood the settlers' intentions', John Hirst sneered, 'there would have been more violence sooner'.[48] And an exasperated McKenna asked what motivated the 'sheer force' and abundance of such engagements with Australia's European settlement: 'Almost every writer feels that they can't understand the country in which they live without first confronting the history of dispossession'.[49] In agreement with Agnew's remarks on reenactment's presentism, I am less interested in proving the historical impossibility of Grenville's fantasy of peaceful conquest and consensual dispossession than paying attention instead to the redemptive narrative the author constructs around her fantasy of going back in time and stopping the brutality the second time round. It is here that the similarities to *Outback House*'s narrative about the colonial past are most striking.

When conceived within the Christian imagery of guilt and redemption, a peculiar characteristic of such imaginary returns to the colonial past comes into relief: their reliance on the 'confessional mode of testimony'.[50] For Grenville, confession and the desire for absolution are central concerns. *The Secret River*'s protagonist is modelled on the author's ancestor and she repeatedly 'confessed' both her fear of finding out about Wiseman's likely participation in acts of dispossession and murder, as well as her sense of an inherited settler-guilt. This reliance on the confessional mode, moreover, engenders a particular conception of how to adequately address this past. Here Grenville, in another of her contentious metaphors, invokes the process of knitting: 'you ... go back to the point where it went wrong ... and you unpick it, like knitting, you've got to undo the whole thing, all those 20 rows, to where you went wrong, where you dropped the stitch, work out why you dropped it, what your choices were, and then you go on'.[51] She thus takes some originary moment as holding the explanatory key[52] to all that has succeeded it and conceives of the return to 'where it went wrong' as a means of 'undo[ing] the whole thing': to 'set right' past wrongs, to resolve settler guilt and to enable a renewed sense of belonging. But Grenville's fantasy of time-travel hinges on a paradox since her desire to pick up the stitches comes up against the factual past *wie es wirklich gewesen* (Ranke). As John Hirst has remarked (drawing attention to Grenville's use of the past conditional), 'Grenville wants there to have been peace, but she knows there was war'.[53] While the author wants the narrative of the nation's past to bestow identity and unify, literary critic Eleanor Collins expands, 'the stories of first contact with which white Australian history must begin are almost always stories of division: of misunderstanding and fear, of brutality and suffering'.[54]

Significantly the confessional is also a distinguishing feature of television reenactment. In *Outback House* it took the explicit form of 'video diary cameras' used by all participants to 'let us into their private thoughts about life on Oxley Downs'[55] and was evoked in numerous other instances in which Indigenous and non-Indigenous participants were invited to reflect on their relationship to the past they were performing. In her reading of this show, Catriona Elder furthermore detects a fantasy of returning to the moment 'where it went wrong' similar to Grenville's to motivate *Outback House*'s re-staging of nineteenth-century outback life. She understands such narratives to answer and counteract recent challenges to non-Indigenous identity in Australia. The protests which disrupted the Bicentenary celebrations in 1988, the High Court decisions to recognise native title in the case of Mabo in 1992 and Wik in 1994, the impact of the 1997 'Bringing Them Home Report' that documented the forcible removal of indigenous chilren, as well as the disconcerting effects of the 'history wars' have 'unsettled non-Indigenous people's certainty about belonging'.[56] While Elder observes a growing awareness and recognition of Indigenous injury and unjust dispossession, she also identifies a simultaneous, redemptive, 'process of refixing' of non-Indigenous senses of belonging in which *Outback House* participates. In ways that echo Andreas Huyssen's comprehension of the current memory boom as expressing a desire for 'spatial and temporal anchoring', Elders understands the series as trying to re-imagine a 'fixed' connection between non-Indigenous Australians and the land they claim for their nation state.[57] However, like Grenville, the programme's narrative cannot relate its redemptive story without coming up against instances of Australia's history that do not assimilate easily to this agenda.

In *Outback House,* this flawed attempt to fix the past by acknowledging a shameful past while simultaneously affirming a settler sense of belonging played themselves out most prominently in the programme's peculiarly displaced performances of the moment of colonial beginning.

2.2.1 Stuttering beginnings

The first of these curious inceptions occurs as soon as the re-enactors have changed into period attire. At the outset, only Paul Allcorn and his male co-workers travel to Oxley Downs while the women and children stay behind. When they 'enter' the past three days later, they are escorted by Malcolm Burns, a Wiradjeri man signed on as Indigenous station hand, who, as the voiceover informs us, 'has stayed behind to accompany the women on the journey through his traditional land'.[58] Unlike his ancestors, whose historical role he is to re-enact, Malcolm thus is a latecomer to *Outback House*'s colonial pastoral. When Paul and the men perform their arrival in 1861, the farmland remains devoid of a – potentially disruptive and unsettling – Indigenous presence. They act as owners and hosts in welcoming Malcolm and the women to the station.

Malcolm, in turn, can only welcome other latecomers – both to the programme and to Australian history. In episode four Russell, a new volunteer, arrives together with Singh, cast in the role of an Indian hawker, who sells his goods to outback squatters. To these outsiders and, here, the non-British transitory vendor in particular, Malcolm enacts a ceremony of Indigenous 'Welcome to Country'.[59] The Allcorns and their non-Indigenous workforce, on the other hand, participate in this ritual of beginning not as guests but as hosts.

Fittingly therefore, the Anglo-Australian Russell, who has signed on as the new station hand, seeks to move beyond his status as a newcomer and attain an equally secure sense of belonging: within minutes after his arrival on the station, he announces that 'I want to be buried here!'[60] In the context of the programme's interest in affirming a non-Indigenous sense of being at home in the Australian landscape, Russell's curious desire to inter his twenty-first-century bones in the simulated nineteenth-century soil of Oxley Downs might be read as a particularly strong case of what Elder describes as 'refixing'. It also curiously participates and updates a nineteenth-century literary mode of addressing the moment of colonial beginning. Desiring to assert a sense of belonging to the land and to demarcate materially the origins of settlement, Victorian literature from Australia abounded in images of the lonely grave. In these texts, the act of claiming the land and the ceremony of burying those who lost their lives in the process become, in the words of Ian Reid, 'hardly separable as rituals that link "England's name" with its new possession'.[61] Russell, in desiring to place his human remains in the ground of Oxley Downs, might analogously be said to establish the beginning of his own heritage as inseparable from that of the nineteenth-century sheep station, thus firmly rooting himself and his ancestry in Australian soil.

Significantly, none of *Outback House*'s recurrent beginnings approximates the actual moment of first contact, in Grenville's terms, the 'point where it went wrong'. Dispossession has already taken place when the volunteers arrive in 1861, the land has been cleared and a home awaits their occupancy. The station's potentially violent and guilt-inducing history of conflict and dispersal, in turn, is displaced onto an unnamed previous owner. Viewers and re-enactors learn about this pre-history only in mediated form from Oxley Downs' daybook, a journal that Paul finds on his first day on the station. This journal, the voiceover explains, was 'left behind by his bankrupt predecessor [and] spells out the challenge for the new squatter'.[62] With the daybook, the programme thus simultaneously acknowledges and disavows the station's violent past. It shows *Outback House*, like Grenville in her interviews, to be predicated on a discrepant knowledge about this history in wanting 'there to have been peace' but knowing that 'there was war'. It seems that the tension between these two sets of historical knowledge characteristic of the settler colonial perspective on Australia's past can only be sustained and endured in an act of 'active forgetting', a kind of Nietzschean liberation from history's burden.[63] Julie's first impression of the squatter's house might be taken to articulate this 'active forgetting' when she states: 'I didn't think it was going to be so dark', and adds in an optimistic tone, later characterized by the voiceover as 'the true spirit of a squatter's wife from the 1860s',[64] that the place is 'going to be a lovely home'.[65]

It is no coincidence that the structural impossibility of redemptive settler colonial narratives – wanting there to have been peace but knowing there was war – articulates itself around these moments of arrival and beginning. As a juncture 'resonant with legal and ideological implications for colonialist discourse', Peter Hulme asserts, 'it is always this beginning moment that the discourse hesitates over'.[66] *Outback House* therefore not only enacts Australia's settlement but, inadvertently, also reiterates a curious characteristic of colonial historiography. As Hulme has long observed with regard to the repetitive obsessive re-telling of stories of colonial encounter between Britain and the Americas in English literature, 'England's own beginning in the New World consisted of a series of stutterings'.[67] In *Outback House*, the unsettling legal and ideological implications of beginning, as well as the process of 'active forgetting' so typical of twenty-first-century settler colonial discourse that attempts to settle and 're-fix' these ruptures, were expressed most prominently in the process of land selection that the show's last episode re-enacted.

Outback House's producers chose the year 1861 for a highly significant reason. As the voiceover insistently reiterates, the date marks a threshold in New South Wales' regime of land ownership. Whereas farmers could previously lease crown land 'for a pittance' from the colonial government, the authorities now aimed to 'take the land away from the squatters and give it to small farmers called selectors'. As of January 1862, we learn, 'every white Australian had the right to buy crown land for a pound an acre'.[68] Historically, this shift in land title policy was part of a strategy to change settlement patterns and encourage more land-intensive forms of agriculture closer to New South Wales' metropolitan areas. In the programme, the act of selection is understood as constituting the beginning of a twenty-first-century sense of non-Indigenous belonging to the Australian landscape. Within *Outback House*'s presentist agenda of articulating and eventually absolving settler guilt, this beginning is again curiously displaced.

On the one hand, the programme portrays this process as enabling 'small' people to buy into 'the great Australian dream of property and independence'.[69] Within this narrative, 'faith, community spirit and team work', as well as 'shar[ed] ownership' between the squatter and his 'trusted [non-Indigenous] employees' provide the key to success' and enable the volunteers to ward off potential buyers from the outside and hold on to Oxley Downs.[70] On the other hand, this process is related in terms of twenty-first-century concerns with Indigenous dispossession. In this context, selector Paul Allcorn confesses his 'sensitivity to his [Malcolm's] feelings because the process of selection was really about further taking the land from his people'.[71] In a move characteristic of the redemptive narratives of peaceful dispossession developed by Grenville and in *Outback House*, Paul eventually decides to act against the demands of historical accuracy and ask Malcolm to join him and his employees in selecting the land. But Paul's offer to dramatize retrospectively a peace that never was via the construction of a shared beginning on Oxley Downs is declined. As Malcolm knows, there was war:

> I've decided that no, I'm not going to get involved in the actual land legislation because it means that they're going to be putting some final title on the land and it sort of gives them some type of ownership whereas I don't believe that you can actually own the land its, its, I belong to the land. It owns me.[72]

2.2.2 'After the historical comes the conditional'[73]: Television reenactment as alternate history

Interestingly this structure of desired but impossible redemptive narratives finds its equivalent in television reenactment's temporal framing. Historical texts habitually employ the past tense in their representation of earlier events and processes. The dominant temporal modes of television reenactment, by contrast, are the present tense and the aforementioned past conditional. Similar to the evocative power of the Christian imagery of guilt and redemption, these temporalities significantly affect and shape the stories they relate about the past.

If historians employ these unconventional grammatical tenses, they do so deliberately to achieve specific effects in their texts. Donna Merwick's use of the present tense in *Death of a Notary*,[74] for instance, attempts to rupture the notion of hindsight that necessarily informs all historical writing. By means of the present tense, her book disturbs the retrospective teleology that we ascribe to historical events when putting them in a logical, causal order and aspires to return to the seventeenth-century past the immediacy once felt by historical actors, who could yet know the future outcome of their actions.[75] In *Outback House*, the present tense works to a similar effect. Historical events unfold as we watch them. They are brought about within the 'now' by contemporary actors, who make choices and cannot anticipate their future success or failure.

Like the present tense, the past conditional – so characteristic of the voiceover in television reenactments and so dominant in Grenville's articulations of her desire to return to the past – is rarely employed by historians. The use of this temporal mode hinges on an imaginary and silent 'if clause' that generates a temporal simulacrum: *if* the volunteers had lived on Oxley Downs in 1861, then what they *are experiencing now* (present tense) *is just as it would have been* (past conditional) in the past they are performing. In historical writing, probably the best known and well-respected example of such so-called counterfactuals is Robert William Fogel's work on the significance of railroads for economic growth in the US from 1964.[76] In this work, Fogel explores how a hypothetical America without railroads would have been industrialised and measures the effects this lack would have had on the nation's economy. In recent years, counterfactual history has gained new currency with the publication of *Virtual History: Alternate History and Counterfactuals* by neo-conservative historian Niall Ferguson,[77] who regards it as ideological tool against what he calls deterministic theories of history; that is histories posited on 'the existence of a Supreme Being, or Reason, or the Ideal, or the class struggle, or the racial struggle'.[78] In parallel to Merwick's rationale for employing the present tense, Ferguson contends that counterfactuals give back the past its contingency and discourage teleological interpretations endorsed all too easily by our retrospective understanding of the past. Paying attention instead to the contingencies that historical actors felt – the potential futures of their life-worlds and the 'would have beens' of our past – he argues, yields a richer understanding of this past.

Most historians consider Ferguson's theory as problematic not only for its neo-conservative agenda. In his famous 'What is history?'

lectures from 1961, British historian E. H. Carr dismissed all 'what if?' speculation as a mere 'parlour game'. And E. P. Thompson, more damning in his judgement, rejected this kind of conjecture as '*Geschic htswissenschlopff*, unhistorical shit'.[79] Historical writing is necessarily a representation of the past, Keith Jenkins has argued more recently in a crushing critique of Ferguson, and is therefore necessarily tainted by the historian's perspective: 'History necessarily involves "looking back" [thereby] introducing *hindsight* and *anachronism* as a necessary (formal) component of history *per se* ... the whole point of history is *not* to know about past events, etc., as a contemporary witness *might*, but as historians *do*'.[80]

This criticism on the part of professional historians notwithstanding, counterfactual history and its literary counterpart, alternate history,[81] as well as historical reenactment, all endorse and promote the simulacra opened up by the past conditional. And all three genres have experienced significant growth in recent years. Echoing Agnew's observations about the presentist agendas underpinning the *House* format, historian Gavriel Rosenfeld attributes the current popularity of such historiographical games to a postmodern intellectual climate 'hospitable to allohistorical speculation'.[82] He understands allohistorical narratives within the mode of the 'would have been' as 'inherently presentist' and driven by their authors' '[b]iases, fears and wishes, the desire to avoid guilt, the quest for vindication'.[83]

For the purposes of my linked analysis of *Outback House* and the Grenville debate, the discussion around allohistorical narratives yields an interesting perspective on the 'stuttering' re-interpretations of the moment of colonial beginning. As Karen Hellekson explains in her study of the genre, alternate history similarly plays with notions of beginning and is often 'concerned with origin, development or cause', relying on the notion that 'an event in the past caused our present'.[84] In addressing this past incident, she explains, alternate history seeks to introduce a point of divergence to a 'crucial point in history'[85] thereby meaningfully changing the present. This is akin to the rationale underpinning Grenville's as well as *Outback House*'s motivation for returning to the past. Such stories often 'rely on time travel, allowing someone with knowledge of history to pinpoint an event and attempt to sabotage it. By controlling cause, these texts tell us, one can control effect'.[86]

While this is a valid literary strategy, a similar sabotaging of the past in *Outback House* must ultimately clash with Australia's history *wie es wirklich gewesen*. The non-Indigenous vision of a shared perspective on the nation's colonial beginning established in the temporal simulacrum of the 'would have been' can therefore only be of limited success. As soon as Paul expresses his belief in 'community spirit' and voices his wish to make Malcolm one of the selectors, the voiceover intervenes – this time in past tense – and affirms that there was war: 'Only white men had the right to select land'.[87]

2.3 Oxley Down's daybook

As *Outback House*'s final day dawns, an idyllic panorama of station life authenticates the volunteers' arrival in the nineteenth century. Along with these images, we hear Julie Allcorn reading out her last entry to Oxley Downs' daybook. She is speaking 'in character', reporting an ordinary Boxing Day morning, while the pastoral scenes she describes flicker across the screen:

> Boxing Day, 26th of December 1861: up at five thirty am to the noise of Claire having a discussion with Molly [the cow] about entering the kitchen. Bernie is raking the hay out at the front of the house. Hopefully the fire is being lit because I need a coffee quite badly … We need to wash up all the glasses and tidy up the house before we go. The flies are up and it is sticky, so it will be hot. There is so much to do, so I will dress quietly and help clean up. Dan has arrived to help as well.
>
> I shall leave this place with memories that are both happy and sad. This has probably been the most emotionally traumatic roller coaster ride I have ever been on with my children. Maybe next year will bring a new adventure. Happy New Year.[88]

Julie's final New Year wishes invite viewers to imagine the station's journal as staying behind in 1861 when the volunteers leave the programme. Just as it awaited Paul on his first day in the past and provided the re-enactors with the station's prehistory, the book is now addressed to the 'next adventure' in Oxley Downs' 1862 future. In its function of enabling communication between the site's past and present, the daybook thus becomes the highly charged symbol of reenactment historiography. This is so for a number of reasons. Firstly, elaborately written with quill and ink, Julie's entry acquires an aesthetics of pastness that suggests an understanding of her description of Boxing Day 1861 as 'the real story … wrought from experience'. The scene thus reiterates reenactment's privileging and sanctioning of a sensual and corporeal access to the past.

Secondly, the daybook also participates in the temporal doubling that characterised *Outback House*'s historical narrative. On the one hand, Julie expresses her perceptions of Boxing Day morning 1861 in the present tense. Her entry thus affirms notions of an immediate access to the past so characteristic of reenactment. Another temporal mode is established, however, by the emotional pictures of farewell from Oxley Downs that follow Julie's narrative: as the volunteers reluctantly walk up a hill and away from their 1861 home, the station bell suddenly rings out. The group stops and turns around to look back once more at the deserted houses below. We see a close-up of the bell swinging in the wind and hear a door rattling. These images are followed by long, drawn-out shots of well-worn work boots just outside the front door, of kids' toys inside the house and of a vase with fresh flowers on the kitchen table. The imagery suggests an ongoing presence of the station's nineteenth-century inhabitants, who seem to hover just outside the camera frame. The ghostly scenes thus make plausible a fantasy that imagines Oxley Downs as located outside of linear, historical time and as co-present with the now. While the building exists in the twenty-first-century present it is also inhabited by the ghosts of the past.[89] The station is rendered as a site of pure potential within the 're-enactive' time of the 'would have been'.

Thirdly, in the future (that is, our present), we are invited to believe, the daybook and the sensual experiences of the past it records will be the single, authoritative source to remain and document life on Oxley Downs. In the context of Australia's contemporary history wars, the circularity of this process (by which twenty-first-century re-enactors are accorded the authority to record Australia's nineteenth-century history) acquires an ideological dimension. Julie Allcorn's nineteenth-century daybook entry silences the present's conflicting plurality of perspectives and endorses a partial view of the past as true. The daybook thus answers a particular desire for spatial and temporal anchoring and contributes to the contemporary 'process of re-fixing' in confirming a link between non-Indigenous Australians and *their* continent. Similar to Russell's avowed desire 'to be buried here', the journal – in rendering as authentic a non-indigenous twenty-first-century fantasy of the colonial past – conveys this history as always already settled.

In the context of this fraught 're-enactive' perspective on the past, Paul Allcorn's concluding reflection on his experience at Oxley Downs simultaneously makes the case for reenactment's identificatory power as well as inadvertently drawing attention to its dangerous ideological potential. Asked what the experience means to him, he replies: 'It's not just a memory. We are our memory. We are what we experience'.[90]

Notes

1. K. Grenville (2005) *The Secret River* (Melbourne: Text Publishing).
2. K. Grenville and R. Koval (17 July 2005) 'Books and Writing: Interview with Kate Grenville', *Radio National*. Transcript available at: <http://www.abc.net.au/rn/arts/bwriting/stories/s1527708htm>.
3. For an accessible summary of the debate, see S. Macintyre and A. Clark (2003) *The History Wars* (Melbourne: Melbourne University Press).
4. Henry Reynolds quoted in B. Attwood and S. G. Foster (eds) (2003) *Frontier Conflict: The Australian Experience* (Canberra: National Museum of Australia), p. 16.
5. Grenville and Koval, 'Books and Writing'.
6. M. McKenna (1 December 2005) 'Writing the Past: History, Literature and the Public Sphere in Australia'. Public lecture held at the Queensland College of Art, Brisbane. Transcript available at <http://www.humanitieswritingproject.net.au/mckenna.htm>.
7. I. Clendinnen (2006) 'The History Question', *Quarterly Essay*, 23, p. 16.
8. K. Grenville (2007) 'Response: The History Question', *Quarterly Essay*, 25, p. 70.
9. Ibid., p. 72.
10. Clendinnen, 'The History Question', p. 20.
11. J. Hirst (2005) *Sense and Nonsense in Australian History* (Melbourne: Black Inc.), p. 87.
12. *Outback House* (ABC 2005): 8.46.05.
13. Ibid., 8.47.27.
14. T. Hunt (2006) 'Reality, Identity and Empathy: The Changing Face of Social History Television', *Journal of Social History*, 39:3, p. 852.
15. See T. Gronberg (2001) 'Siting the Mordern', *Journal of Contemporary History*, 36:4, p. 687. See also V. Agnew (2004) 'Introduction: What is Re-enactment?', *Criticism*, 46:3, pp. 327–39; A. Cook (2004) 'The Use and Abuse of Historical Re-enactment: Thoughts on Recent Trends in Public History', *Criticism*, 46:3, pp. 487–96; S. Gapps (2002) 'Performing the Past: A Cultural History of Re-enactments' (Ph.D. dissertation, University of Technology, Sydney); I. McCalman (2004) 'The Little Ship of Horrors: Reenacting Extreme History', *Criticism*, 46:3, pp. 477–86.
16. The quotation is taken from the DVD's cover notes. See *Outback House*, 2005.
17. V. Agnew (2007) 'History's Affective Turn: Historical Re-enactment and its Work in the Present', *Rethinking History*, 11:3, pp. 301–2.
18. M. Arrow (2007) '"That History should not have ever been How It was": *The Colony*, *Outback House* and Australian History', *Film & History*, 37:1, pp. 54–66; C. Elder (2007) '*Outback House*: Non-Indigenous Belonging, Colonialism and Reality Television'. Paper presented as part of the seminar series at the Research School of Humanities, Australian National University, Canberra, 14 September 2007.

19. My discussion of *The Secret River* will not be concerned with the novel itself, but will focus exclusively on the public claims that Grenville has made for it and the responses of various historians to these statements.
20. 'Bridging the Divide: History You Can "Taste"', *Manly Daily* (2 August 2005), p. 10.
21. Grenville and Koval,'Books and Writing'.
22. J. Sullivan 'Making a Fiction of History', *Age* (21 October 2006), p. 12. See also Grenville and Koval, 'Books and Writing'.
23. P. Brown 'What Kate Did Next', *Brisbane News* (21 September 2005), p. 10.
24. Sullivan, 'Making a Fiction of History'.
25. J. W. Scott (1991) 'The Evidence of Experience', *Critical Inquiry*, 17:4, pp. 773–97.
26. Sullivan, 'Making a Fiction of History'.
27. This is a reversal of an eighteenth-century conception of spatial distance as temporal: the farther away a place was located from the imperial centre, the more primitive its society was thought in relation to European civilisation. For a similar argument see Agnew, 'History's Affective Turn', p. 304.
28. F. Rokem (2000) *Performing History: Theatrical Representations of the Past in Contemporary Theatre* (Iowa: University of Iowa Press), p. 13.
29. M. Sturken (1997) 'Re-enactment, Fantasy, and the Paranoia of History: Oliver Stone's Docudramas', *History and Theory*, 36:4, p. 68.
30. Together with my colleague Sabine Lucia Müller, I have explored these ideas in more detail in S. L. Müller and A. Schwarz (2008) 'A Ready-Made Set of Ancestors: Re-Enacting a Gendered Past in *The 1900 House*' in A. Schwarz and S. L. Müller (eds) *Iterationen: Geschlecht im kulturellen Gedächtnis* (Stuttgart: Metzler), pp. 89–110.
31. *Outback House* (ABC 2005): 1.08.20.
32. Ibid., 1.08.45–1.09.25.
33. M. Foucault (1977) 'Nietzsche, Genealogy, History' in *Language, Counter-Memory, Practice: Selected Essays and Interviews*, (ed.) Donald F. Bouchard (Ithaca: Cornell University Press), p. 148.
34. *Outback House* (ABC 2005): 1.09.48.
35. Ibid., 1.02.26.
36. This, and all other comments, are available at <http://www.abc.net.au/tv/outbackhouse/>.
37. A list that illustrates the extent of viewers' attention to detail was posted by 'What Rubbish' on 19 June 2005. For instance, one viewer contended that: 'Work staff would not have been allowed to call the squatter and foreman by their Christian names. The cook would have been given full authority to run the kitchen as she saw fit. The workmen would have simply slept in a bunk room, there would have been no dividing sheets forming little cubicles. The workmen would not have been allowed to drink on the property, and staff certainly would not have been lolling in bed at 8:30 in the morning'.
38. I. Baucom (1999) *Out of Place: Englishness, Empire and the Locations of Identity* (Princeton: Princeton University Press), especially pp. 164–89; P. Gilroy (2004) *After Empire: Melancholia or Convivial Culture?* (London: Routledge), especially pp. 95–132.
39. A. Huyssen (2000) 'Present Pasts: Media, Politics, Amnesia', *Public Culture*, 12:1, p. 25.

40. Ibid., p. 31.
41. Ibid., p. 36.
42. See Agnew, 'History's Affective Turn', p. 300.
43. The quotation is taken from the DVD's cover notes. See *Outback House* (ABC 2005).
44. Agnew, 'History's Affective Turn', p. 304.
45. K. Grenville 'First Person', *Guardian* (18 February 2006), p. 3. See also K. Grenville (2006) *Searching for* The Secret River (Melbourne: Text Publishing), especially Chapters 2, 4 and 13.
46. 'Picking Up the Stitches', *Canberra Times* (24 September 2005), p. 12.
47. Grenville and Koval, 'Books and Writing'.
48. Hirst, *Sense and Nonsense in Australian History*, p. 84. See also Clendinnen, 'The History Question', p. 19; McKenna, 'Writing the Past'.
49. McKenna, 'Writing the Past'. Interestingly, this claim extends to include himself and Clendinnen, who have both recently published their own interpretations of the country's colonial past. See also E. Collins (2006) 'Poison in the Flour', *Meanjin*, 65:1, p. 40.
50. McKenna, 'Writing the Past'. Foucault considers confession 'as one of the West's most highly valued techniques for producing truth'. See M. Foucault (1990 [1976]) *The History of Sexuality: An Introduction*, vol. 1 (New York: Vintage Books), p. 59.
51. 'Picking Up the Stitches', *Canberra Times*.
52. Collins, 'Poison in the Flour', p. 40. See also S. Kossew (2007) 'Voicing the "Great Australian Silence": Kate Grenville's Narrative of Settlement in *The Secret River*', *The Journal of Commonwealth Literature*, 42:7, p. 17.
53. Hirst, *Sense and Nonsense in Australian History*, p. 85.
54. Collins 'Poison in the Flour', p. 40. For a similar argument see G. Hage (2002) 'Polluting Memories: Migration and Colonial Responsibility in Australia' in M. Morris and B. de Bary (eds) *'Race' Panic and the Memory of Migration* (Hong Kong: Hong Kong University Press), pp. 349–50.
55. *Outback House* (ABC 2005): 1.21.10.
56. Elder, '*Outback House*: Non-Indigenous Belonging, Colonialism and Reality Television'.
57. In developing the notion of 're-fixing', Elder draws on A. Cerwonka (2004) *Native to the Nation: Disciplining Landscapes and Bodies in Australia* (Minneapolis: University of Minnesota Press), p. 3.
58. *Outback House* (ABC 2005): 1.48.50.
59. Ibid., 4.08.10.
60. Ibid., 4.08:48.
61. I. Reid (2002) 'Marking the Unmarked: An Epitaphic Preoccupation in Nineteenth-Century Australian Poetry', *Victorian Poetry*, 40:1, p. 10.
62. *Outback House* (ABC 2005): 1.31.00.
63. See F. Nietzsche (1990) 'History in the Service and Disservice of Life' in *Unmodern Observations*, (ed.) W. Arrowsmith (New Haven: Yale University Press), pp. 87–145.
64. *Outback House* (ABC 2005): 4.02.18.
65. Ibid., 1.53.20.
66. P. Hulme (1986) *Colonial Encounters. Europe and the Native Caribbean 1492–1797* (London; New York: Methuen), p. 193.

67. Ibid., p. 89.
68. *Outback House* (ABC 2005): 8.15.00.
69. Ibid., 8.17.48.
70. Ibid., 8.14.40–8.15.38.
71. Ibid., 8.19.15.
72. Ibid., 8.18.22–8.18.47.
73. C. P. Freund (1999) 'Spilt History', *Reason*, 31:7, p. 50.
74. D. Merwick (1999) *Death of a Notary: Conquest and Change in Colonial New York* (Ithaca: Cornell University Press).
75. For a review of Merwick's book that supports this reading, see M. B. Norton (2000) 'Review: *Death of a Notary: Conquest and Change in Colonial New York*', *The William and Mary Quarterly*, 57: 4, pp. 852–55.
76. R. W. Fogel (1964) *Railroads and American Economic Growth: Essays in Econometric History* (Baltimore: John Hopkins Press).
77. N. Ferguson (1999) *Virtual History: Alternatives and Counterfactuals* (New York: Basic Books).
78. Ferguson quoted in K. Jenkins (2001) 'Review: Niall Ferguson, (ed.) *Virtual History: Alternatives and Counterfactuals*. London: Picador, 1997', *Rethinking History*, 5:1, p. 168.
79. E. P. Thompson (1978) *The Poverty of Theory and Other Essays* (London: Merlin Press), p. 100.
80. Jenkins, 'Review', p. 171.
81. As Gavriel Rosenfeld explains, alternate history traces its roots back to the first allohistorical novels published in post-Napoleonic France. Like counterfactual history and re-enactments, the genre became increasingly popular in the decades after 1960. See G. Rosenfeld (2002) 'Why Do We Ask "What If?": Reflections on the Functions of Alternate History', *History and Theory*, 41, p. 92.
82. Ibid.
83. Ibid., p. 93.
84. K. Hellekson (2001) *The Alternate History: Refiguring Historical Time* (Kent: Kent State University Press), p. 3.
85. Ibid., p. 5.
86. Ibid., p. 34. See also K. Hellekson (2000) 'Towards a Taxonomy of the Alternate History Genre', *Extrapolations*, 41, pp. 248–56.
87. *Outback House* (ABC 2005): 8.19.10.
88. Ibid., 8.40.41–8.41.46.
89. I discuss the employment of the gothic mode in another television re-enactment in A. Schwarz (2007) '"Not This Year!" Reenacting Contested Pasts Aboard *The Ship*', *Rethinking History*, 11:3, pp. 427–46.
90. *Outback House* (ABC 2005): 8.54.28.

3

'Recreating Chaos': Jeremy Deller's *The Battle of Orgreave*

Katie Kitamura

In 2001, British artist Jeremy Deller staged *The Battle of Orgreave*, a partial reenactment of a historical confrontation between striking miners and police. Recreating a violent standoff that took place some 17 years earlier in the same part of South Yorkshire, *The Battle of Orgreave* referenced an event that came to epitomize the ongoing struggle between workers' unions and the Thatcher government of the 1980s.

At the centre of a conflict that polarized the British nation was a single, economically devastated community. Focusing on this community's past and continuing experience, Jeremy Deller's reenactment tracked the social and political resonance of the strike, riot, and subsequent police retaliation. Involving over 800 participants drawn from 20 reenactment societies across Britain, as well as former miners and police officers themselves involved in the original strike, *The Battle of Orgreave* was an immaculately orchestrated piece of theatre, and remains Deller's most ambitious work to date.

In its entirety, *The Battle of Orgreave* inhabited a number of manifestations: a traditional reenactment event, a broadcast television documentary,[1] a social gathering complete with beer tents, and a highly nuanced work of conceptual art. Rather than servicing a single function or speaking the language of a single discourse, *The Battle of Orgreave* was relentlessly proliferating. Often much more than an artwork, Deller's reenactment was, even in the ever-widening language of public and performance art, determinedly original. That self-conscious multiplicity is perhaps characteristic of what French critic and curator Nicolas Bourriaud has termed 'relational aesthetics', by which he means, 'an art taking as its theoretical horizon the realm of human interactions and its social context, rather than the assertion of an independent and *private* symbolic space.'[2]

Relational aesthetics characterizes the work of a number of international artists who first rose to prominence during the nineties; these artists include Liam Gillick, Pierre Huyghe, Felix Gonzalez-Torres, and Vanessa Beecroft. Today, Jeremy Deller's art remains informed by that particular context. His work is first and foremost concerned with human interaction and reaction, and the broader social context reflected therein; nearly all his work can be characterized as an exploration of the ways in which a selected demographic interacts with a suggested premise.

His art, in this sense, can be described as one of selection and suggestion, rather than concrete authorship; Deller is regularly credited as the initiator of an artwork rather than its author. Central to the relational aesthetics movement is the interrogation of aesthetic and authorial control. The art falling into the discourse of relational aesthetics often takes the shape of a performance, event, or intervention involving members of a designated public. Although these events are initiated and controlled by the artist and by the language of art production (and in this sense *The Battle of Orgreave* can be posited as an experiment in organized – even enforced – collective memory), they are also artworks ruled by unpredictability.

Relational aesthetics inhabits the space of the unpredicted and the unforeseen. That space is one that designates itself in temporal terms: in the time between the conception, initiation, realization and then summation of the event, in the very durational nature of the works themselves. It is across the temporal plane that the interest of these works lies. If a conventional work of art can be said to occupy a spatial zone, then the aim of relational aesthetics is to explore durational rather than physical spaces.

For this reason, *The Battle of Orgreave* was a work of relational aesthetics *par excellence.* It is explicitly a work 'outside' the gallery, and by engaging with history, it matches the literal interest in time with a heavily conceptual one. Perhaps most importantly, it was a work that remained forcibly open – not only because of the chaotic nature of its subject matter, but also because of its own unresolved place in the continuing history of the original miners' strike itself.

The Battle of Orgreave functioned as an 'event' in the most literal sense of the word, both in terms of the historical and political reference of the reenactment and the sheer crowd spectacle of the hundreds of participants involved. Further, it was an event that schematically communicated the unresolved tension between the impulse towards uncontainment and the inclination towards control that is finally the central paradox of relational aesthetics.

If *The Battle of Orgreave* was first and foremost about the depiction of chaos, then it was also about the containing of disorder – politically, historically, and, of course, physically. It was, after all, not simply a reenactment of the miners' strike, but a reenactment of the strike and subsequent police retaliation. And if, in the chaotic upheaval of the riot and the ideological challenge to authority it represented, Deller's interest in subverting notions of authorial control was aptly represented, then it seems important to note that this uncontainment was simultaneously channelled into what might be considered a formal, aesthetic, interest in the problem of spillage.

Deller's choice of event is also important here; in choosing to stage a reenactment, Deller conceived of a type of event that would, even outside the discourse of art production, carry with it the tension between containment and uncontainment. The reenactment event inhabits an uncomfortable territory, at once a wholly staged and utterly genuine thing. Simultaneously denatured and potent, it is precisely this contradiction that was of interest to Deller. Of the reenactment process he writes: 'As an artist, I was interested in how far an idea could be taken, especially one that is on the face of it a contradiction in terms, "a recreation of something that was essentially chaos"'. The risks inherent in the depiction of chaos and its own simultaneous containment are, in this sense, central to Deller's artistic project. On a formal level, *The Battle of Orgreave* therefore actively engaged with two bipolar risks: the illegibility of an unmitigated disorder, and the sterility of a denatured chaos. The persistent sense of these two twinned risks, was at the centre of the enduring tension that characterizes Deller's work, here and elsewhere. The artistry of *The Battle of Orgreave* was perhaps not so much in the technical minutiae of the reenactment, perhaps not even in the conception of the project itself, but rather in the conceptual flexibility, the formal laxity, that allowed these various contradictions and paradoxes space to coexist, if only momentarily.[3]

Contradiction and polarity are perhaps at the heart of *The Battle of Orgreave*'s artistic ambitions. Deller claimed his interest in reenactment was essentially an interest in the paradoxical phrase 'living history',[4] a term often employed to describe reenactment events. He was in a sense interested in interrogating and reappropriating that phrase. *The Battle of Orgreave* was fundamentally conceived as a bifurcated work; in it, representations of the living and the historical found two separate conceptual and literal manifestations.

The division between these two aspects – while not necessarily mutually exclusive – was itself schematically expressed in the demographic of the reenactment participants. Representing the 'living' aspect, there were the 'former miners (and a few former policemen) who were reliving events from 1984 that they themselves took part in'.[5] Representing the 'history' element, there were 'members of Battle reenactment societies from all over the country ... well trained in recreating combat and in obeying orders.'[6]

Between the original miners and the reenactment participants, an immediate fissure, an inconsistency in the meaning of experience, was declared. Although the two groups worked together and occupied the same physical space, although they were part of the same 'coming' together, the specific nature of their experience was radically different on nearly every point. These were differences of personal versus collective memory, traumatic repetition as opposed to evocative iteration, fundamental differences in the very nature of the 'event' – or indeed, 'non-event' – in which these two groups were taking part.

I would like to suggest that as an artwork, *The Battle of Orgreave* is animated by a fundamental tension between the evocation of what is, on the one hand, the obliterating totality of the 'real' event, and on the other the structuring mechanism of the 'non-event,' the event that was expected, foreseen, and predictable in the function it fulfills.[7] It is in the inconsistencies between these two very separate notions of eventhood that *The Battle of Orgreave*'s motif of uncontainment versus containment finds its most persuasive expression.

Although the reenactment was certainly an event of its own standing, it is perhaps useful to momentarily consider the reenactment – both 'versions' of it – as part of the after-effect of the still-unfolding event of the original strike and conflict, particularly in light of the fact that what is being explored in *The Battle of Orgreave* is, at least partially, the process of historicization. In *Philosophy in a Time of Terror*, Derrida declares, 'The event is made up of the "thing" itself (that which happens or comes) and the impression (itself at once "spontaneous" and "controlled") that is given, left, or made by the so-called thing'.[8] In this sense, the reenactment can be seen to be composed of a schematic representation of these two separate qualities of the event's after-effect, 'at once "spontaneous" and "controlled"'.

Within the very structure of the event's after-effect, there is thus, on the one hand, an impulse towards the continuation, the perpetual distending of the event (as represented by the element that is 'spontaneous', and which continues to communicate the 'real' of the original event), and, on the other, an impulse towards closure (as represented by the inclination towards 'control'). What finally characterizes an event is not simply its original unforeseen and irruptive quality, but precisely this enduring tension between opening and closure, between the continued evocation of the 'real' event and the more rigid declaration of the 'non-event'. The historical event endures through that very tension, through the paradoxical combination of continued viability and growing historical legibility.

In this context, the reenactment's inclusion of the original miners, drawn into a systematic repetition of a moment of fundamental trauma, clearly reads as something like the 'spontaneous', open element of the after-effect, as that which communicates some continuing aspect of the 'real' of the original event. If the 'real' event is fundamentally disruptive and traumatic, existing beyond the parameters of context (in describing what would constitute a 'major event', Derrida, in an interview discussing the September 11 terrorist attacks, says, 'A major event should be so unforeseeable and irruptive that it disturbs even the horizon of the concept or essence on the basis of which we believe we recognize an event *as such*'),[9] then it can produce no cogent narrative; it proposes no recovery from the trauma of its own 'irruptive' unforeseeability.

The reenactment experience of the miners therefore represented a prolongation of the traumatic, illegible aspect of the original riot. Deller himself describes the strike as '*traumatically* divisive',[10] and declares that these participants were '*reliving* events from 1984 that they themselves took part in'.[11] There was, in a sense, no real surplus of emotion here, but rather the reinforced sterility of a traumatic repetition from which no clear narrative could be drawn.

Here it should be emphasized that the reenactment of the clash was always only a 'partial reenactment'. The emphasis on 'partial' is crucial here. The reenactment did not serve to represent the entirety of the past event through a partial reenactment; instead, it acted as the literal repetition of a fragment, divorced from ideas of representation. It was, however, partial in a very literal sense, in that only a carefully sectioned portion of the day's events was re-enacted, but it was also partial in a looser, more metaphorical sense; partial in the sense of an incomplete statement, a narrative gesture that was unfinished, and thus unrecoverable. The very fragmented nature of the reenactment, however much a product of practical necessity, fitted into the model of a traumatic repetition. Also worth mentioning in this context are the miners' songs, chants, and mantras that were repeated throughout the reenactment and that, as much as they could in another framework represent a kind of nostalgic, sentimental excess of emotion, in this particular light became yet another example of a stuttering, traumatic utterance, reduced to the level of a literal nonsense.

In direct contrast, the presence of the reenactment societies spoke of a very different form of repetition, and thus a very different aspect of the event's after-effect. This repetition, rather than distending the irruptive unpredictability of the original event, functioned in a manner that was essentially predictable, foreseen and foreknown. The reenactment, as embodied by the reenactment societies, was carefully controlled and monitored, an art event that functioned according to a professional timetable, arriving and ending on schedule and per expectation. That status as a 'non-event' becomes almost comically clear when one considers the way in which, on a purely pragmatic level, *The Battle of Orgreave* was deliberately and necessarily concerned with ideas of control. In the notes distributed to reenactment participants, among the 'Conditions of Participation' was the declaration: 'These are by necessity quite strict. We will require a professional attitude and a willingness to stick to an agreed script. You will need to conform to a dress code and to look the part'. Those issues of conformity were, again by necessity, intensified by the process of filming and editing that was integral to the production of the documentary film.[12] It was precisely that portion of the event's impression that was linked to the notion of 'control'.

The presence of the professional reenactment societies in a sense prematurely 'closed' the openness of the original event, ratifying instead the very historicity of the strike. Rather than disturbance and interruption, this instance of repetition functioned as a form of closure, control, and categorization. That element of closure was in a sense inevitable, given that Deller's interest in these societies arose out of his desire for 'the reenactment of The Battle of Orgreave to become part of the lineage of decisive battles in English History'.[13] Deller's very interest in reenactment arose out of its capacity to evoke the continuous weave of history; his relationship to reenactment as a social practice emerged out of an interest in the particularities of the (already established) historical and cultural narrative it evoked.

This historical narrative was perhaps inevitably nostalgic in its nature. In a sense in was the very excess of the reenactment's detail, the technical precision and accuracy of the repetition that was so conducive to the sentiment of nostalgia. Deller himself explicitly denied he was interested in a 'nostalgic interpretation of the strike';[14] however, it seems obvious that nostalgia was critical to the historicizing purposes of the reenactment. In this context, nostalgia was acutely functional, creating a cogent narrative through which the release of emotion associated with the reenactment event could be organized.

If, in this aspect of *The Battle of Orgreave*, nostalgia came rushing back in, then it was present in the very styling and detailing of the event. It could be located in the nostalgic badges, costumes and songs that were easily placed in the context of an overarching British nostalgia for the heroic struggles of the eighties. Perhaps more importantly, nostalgia was crucial in emphasizing the process of historicization that Deller is explicitly interested in; nostalgia helped ground the event firmly in the past. And finally, nostalgia not only contained, but also amplified the emotion of the reenactment, through its very sentimental power.

And if, in a sense, nostalgia always circles around absence, then it is clear that the reenactment societies were animated by a perfect structure of absence, nostalgia, sentiment and the language of historicity. Drawing on a collective notion of the original event – an idea already conditioned by the heavily sensationalized media coverage in 1984 and 1985 – the reenactment societies recreated a sentimental version of a distantly perceived event that was itself only understood through the filter of sentiment. Crucially Deller selected reenactment societies 'from all over the country', and not only the regions directly involved in the strike. They created a highly affective repetition that made no rupture with the present summary of the past; their reenactment was perhaps not interested in recreating the original event, so much as sustaining its social traces.

It is through this nostalgic reenactment that the tension between containment and uncontainment which underpinned *The Battle of Orgreave* reached a contrasting laxity, and something like a sense of loose melancholia. The quality of this melancholia – if indeed melancholia it was – contrasted with both the rigidity of containment, and the overpowering rush of uncontainment. Finally, it was perhaps that peculiar, almost elegiac quality of emotion that altered the pacing of the artwork from that of a mere conceptual exercise.

Perhaps inherent to every gesture of reenactment is the fact of its own failure: the melancholy of reenactment only emerges in light of a supposed loss, in the face of a perceived breakdown. Here the failure to fully recreate the past was placed in high relief by the inclusion of the 'original' miners, the 'genuine' participants of the strike. It was in the failed recapitulation of the past that the genuine emotional power of the nostalgic reenactment was perceived. If Deller himself had something of an ideological aversion to nostalgia, there is nonetheless in his work an *aesthetic* propulsion towards nostalgia, towards its capacity to invoke the sentiment and melancholy that is so crucial to the appeal of his work; much of Deller's work is located in the cultural history of the recent past, and the territory of childhood and adolescence.

Deller's primary interest in the reenactment was always to track the process of history, and I would suggest that through the shared presence of the original miners and the reenactment societies, Deller was able to negotiate a difficult antinomy. In *The Battle of Orgreave*, Deller maintained the porous structure of the authentic event while simultaneously placing it in a historical context; the work presented a legible narrative of nostalgia while simultaneously criticizing and identifying the lapses of that very sentimentality. In capturing the contrasting complexities of the event's after effect, he, in this way, also represented the complicated manner in which an event continues to alter and change, even as it slowly enters into the narrative of historicity.

Nonetheless the apparatus of the work eventually turned to the question of how to frame, and in a sense neutralize, the paradox of this tension. And crucially, although sustaining the paradoxes of 'living history' was central to the complexity of the artwork as event, the gesture towards historical contextualization became, in the aftermath of the aftermath of the event, finally the more dominant – perhaps emphasizing that the borrowed term 'living history' was always in the first instance used to describe historical reenactment societies. In naming the publication that accompanied the reenactment *The English Civil War Part II*, Deller lent the reenactment just one more added sense of closure. Perhaps the process of depicting 'living history' could only ever be sustained by the particularities of the live event itself.

Deller's simultaneous use of the reenactment societies and the original protesters also illustrates an interest – even an anxiety – with regard to the role of authenticity in reenactment. In the doubling up of technical accuracy against experiential authenticity, Deller identified a crucial point of interest in all reenactment events: the role of fakes and false reproductions, and the manner in which authenticity and fakes increasingly keep shared company.

Because of the way Deller structured *The Battle of Orgreave*, the participation of the reenactment societies – and therefore the entire language of reenactment culture – was perhaps necessarily placed in something of a false position in relation to the more obvious, and in some ways more persuasive, authenticity of the original protestors and policemen. But the very fact of their inclusion, alongside the central role they performed in invoking an authoritative historical narrative, perhaps indicates the crucial role of fakes in our relationship to historical narrative.

Fakes grant access to the inaccessible, provide a mechanism for the sorting and cataloguing of the unsortable. In producing a simulacrum of authenticity, they ratify a given historical narrative, even as they heighten the effect of that narrative's authority. The comparative sterility of fakes is in a sense crucial here; they produce a version of the real that is containable, non-irruptive. Nor can this seeming artifice be identified as a strictly contemporary phenomenon; the relative sterility of fakes is perhaps crucial to our basic capacity to integrate and maintain a historical perspective. The overwhelming obliteration of the 'real' event is not, perhaps, fundamentally conducive to supporting the continuity of a historical narrative.

The difficulty of maintaining both the authority of the fake and the *jouissance* of the authentic might, perhaps, be addressed through the very troubled notion of authorial signature – a notion that is always, in a sense, hovering over Deller's work. Carefully tracking and then recording the outcomes of these various interactions, Deller has long attempted to rewrite the identity of the artist as part initiator, part archivist. Examples of this freehand authorship include his *The Uses of Literacy* (1997) project, in which Deller invited (the mostly teenage) fans of the *Manic Street Preachers* to contribute art works inspired by the band's music. The project culminated in an exhibition exclusively made up of other people's work, in which the assembled works veered wildly from depressive poems to exuberant cartoons and apparently non-ironic life drawings. When the piece was eventually purchased by an American collector, Deller dispatched payments to all the contributors – a gesture that seemed as much a part of the project's intent as the original collation itself.[15] However, much as Deller might ostensibly avoid the closure of signature, it seems tenable that the continued tension between the fake and the authentic in *The Battle of Orgreave* was, paradoxically, sustained by the enduring power of his signature, and the injunction to come and witness that is issued through his identity as artist.

Central to the reenactment event is the notion of the 'coming' – the *jouissance* of the event that always 'comes'. This is implicit to the idea of the 'real' event, and if the 'real' event is always coming, then it is also always issuing an invitation to come, and an imperative to witness. According to Nicholas Royle, that coming is always addressed to a spectral other that is central to our notion of history; in this sense, the 'coming' is perhaps the point at which the overflow of the 'real' event brushes up against a continuous idea of history.[16]

In this 'coming' there is a notion of historical continuity and an idea of authority that is reinforced by the artist's signature. And it can surely be no mere coincidence that so much of Deller's work as an artist is preoccupied with the injunction of a very literal coming, a drawing together of people. Whether it is the staging of a group bus tour commemorating the hundredth anniversary of an arbitrary picture postcard (*Constable, Can't You See My Predicament*, 1994) or the dissemination of mock calling cards (imprinted with a variety of messages, ranging from 'We really mustn't meet up and go for a drink some time' and 'Please help me, I went to a single sex public school' to 'A rich seam of melancholy' (*Assorted Calling Cards*, 1992–5), Deller is clearly and literally interested not simply in the gesture of instigation, but more specifically in the role of summoning.

That gathering of people is central to the notion of relational aesthetics, and to the idea of artistic command and authorial signature; Bourriaud writes: 'The artwork is thus no longer presented to be consumed within a "monumental" time frame and open for a universal public; rather, it elapses within a factual time, for an audience *summoned by* the artist'.[17] The passage of that audience is crucial to the form of relational aesthetics ('What human *flow*, governed by what forms, thus passes into art forms?'[18]), and it is perhaps in this coming that the *jouissance* of the event is paradoxically contained by both historical narrative and aesthetic form.

And, particularly with reference to Royle's notion of addressing a spectral other, it seems equally important to note that *The Battle of Orgreave* was an artwork that was thoroughly witnessed by an audience drawn from all across Britain – and that this audience, the physical manifestation of the appeal to 'come' that is made to the spectral other, watched the event from a god's eye view, lodged in the hills of Orgreave.

Notes

1. Directed by Mike Figgis and co-produced by Artangel and Channel 4.
2. N. Bourriaud (2002) Relational Aesthetics, trans. S. Pleasance and F. Woods, with the participation of M. Copeland (Dijon: Les presses du reel), p. 14.
3. J. Deller (2002) 'Introduction', in *The English Civil War Part II: Personal Accounts of the 1984–85 Miners' Strike* (London: Artangel Publishing), p. 7.
4. Deller, *The English Civil War Part II*, p. 7.
5. Ibid.
6. Ibid.
7. While in a sense, every event is an event, and every event carries with it some element of unpredictability, Derrida seems to evoke an idea of greater and lesser states of eventhood, as in relation to the September 11 terrorist attacks and his description of what might constitute a 'major event'. See G. Borradori (2003) *Philosophy in a Time of Terror, Dialogues with Jurgen Habermas and Jacques Derrida* (Chicago: Chicago University Press).

8. Ibid., p. 34.
9. Ibid., p. 33.
10. Deller, *The English Civil War Part II*, p. 7 (emphasis added).
11. Ibid. (emphasis added).
12. Deller, Conditions of Participation, n.p.
13. Deller, *The English Civil War Part II*, p. 7.
14. Ibid.
15. In his equally playful *Acid Brass* (1997–8), Deller organized a series of concerts demonstrating a new musical hybrid of traditional brass bands performing acid house dance anthems.
16. See N. Royle (2003) *The Uncanny* (New York: Routledge).
17. Bourriaud, *Relational Aesthetics*, p. 29.
18. Ibid., p. 74.

4
On Being a Mobile Monument: Historical Reenactments and Commemorations

Stephen Gapps

A few years ago I was at an American Civil War reenactment 'talking authenticity' – as re-enactors do – with someone who claimed to be re-enacting her great-grandmother. She spoke to me about the benefits of researching a 'real, historical person' instead of portraying a generic character of 'someone from the Civil War'. She firmly believed an historical person could be quite accurately portrayed in both costume and character, particularly if there were surviving photographs, diaries, or letters. She also recommended researching such things as 'local newspapers of the times' to gain a contextual understanding of the events and issues 'your character would have known about'. In her case, she could re-enact an ancestor. For others without this connection, she recommended choosing an 'unknown' historical figure and researching them in detail. Then she proudly said to me that she felt as though she were a 'mobile monument' to her great-grandmother's memory.[1]

The idea that re-enactors could be 'mobile monuments' is intriguing. It reminded me of ideas about the 'self as museum' – a suggestion that shaky postmodern subjectivities might somehow be stabilized by aligning the self with the archive. And this was certainly quite a literal and a very public expression of an individual aligned with an archive. She was not just writing her family history, but play-acting it out in front of other re-enactors and their audiences in an officially sanctioned commemoration.

In fact, many re-enactors are walking, talking, 'mobile museums'. Their car-boots and trailers are full of painstakingly recreated historical artefacts that they 'bring to life' at reenactment events. Why can't they be monuments as well? Rather than visiting a gravestone or compiling her family tree, our re-enactor was in effect creating a monument to her ancestor for use in events that operate as open-air stages of national remembrance – historical reenactments (Figure 4.1).

Figure 4.1 Two versions of William Bligh

There has been increasing academic and media scrutiny of the political alignment of unsanctioned, nostalgic battle reenactments with official commemoration of the Civil War in the US.[2] Often, these 'weekend warrior' reenactments give public expression to otherwise politically questionable histories, such as the 'lost cause' of the Confederacy. They also give excessive weight to the military *as* history; a history that is widely held as unquestionably sacred.

For this re-enactor, historical accuracy in representation is obviously a critical component of her practice. Despite perhaps being driven by a nostalgia for a nobler age or for personal connection to these grand and sacred narratives of history, any attention to the byword of reenactment – authenticity – demands a widening of the symbolic and representative repertoire outside the usual focus on the soldier in battle. In effect, as the 'soldiers' who have historically driven the practice of reenactment become increasingly attuned to issues of authenticity, then so too the usual boundaries of, for example, Civil War reenactment must expand to include women, civilians and slaves.

Over the last 20 years or so the presentation of authentic histories has become a central tenet of reenactments. Authenticity is a mark of performance competency among re-enactors – it divides the serious from the not-so-serious. Because of this, the subcultures and practices of amateur re-enactors have also become much more complex affairs than merely escapist costumed play-acting. Re-enactors increasingly negotiate questions of historical representation and method.

This essay investigates what this might mean for historical commemorations that incorporate reenactments. It outlines the public reception of two quite different commemorative responses to the 1804 Australian 'Battle of Vinegar Hill' – a series of monuments and a reenactment. In the context of an increasing level of official sanction of amateur historical re-enactors at commemorative events, I ask what might happen if, as our re-enactor suggested, we think of re-enactors as 'mobile monuments'?

But first, I must declare a conflict of interests here. I have been involved in designing, coordinating, and participating in historical reenactments for over ten years. I have fought at the Battle of Hastings and marched 137 miles to the Battle of Gettysburg. I have escorted Admiral Nelson and Lady Hamilton through the streets of Portsmouth before the Battle of Trafalgar. I have lived (albeit briefly) in Viking long-houses, camped with English Civil War armies and cooked with Republican Romans. Closer to home, I have been a nasty British redcoat arresting Irish convicts, but have also been an Irish rebel at the Battle of Vinegar Hill.

Throughout all these battles and marches I have gone to some extreme lengths in pursuit of the 'holy grail' of authenticity. I have made leather shoes by hand from a pattern based on a tenth-century archaeological find and hand sewn a medieval tunic. I have worn out two pairs of shoes in a week by marching over a 100 miles. And recently I even sported a 'mullet' style haircut to imitate the Saxons depicted on the Bayeaux Tapestry.

As both a public historian and self-confessed re-enactor, I have gained some interesting insights into reenactments and the practices of re-enactors. I have found reenactment to be much riskier work than formal historical writing. It is quite nerve racking, wearing the contents of your research as costume. Very often nowadays, the successful public reception of a reenactment is measured in terms of the historical accuracy of re-enactors' costume and props. Re-enactors' reputations as walking, talking archives increasingly precedes them.

I often find myself, just like our 'mobile monument', constantly striving for a more historically accurate presentation. My 'living history' aesthetics are now very much piqued when, for example, I see someone at a reenactment with a mobile phone or wearing modern glasses. Anachronisms disturb my need to uphold the status of reenactment as history work and I see myself in danger of becoming an authenticity fetishist, or as fellow re-enactors say, an 'authenticity fascist'.

Since the 1970s, historical reenactment has come to encompass a very broad range of 'performance history'. It is now a byword for almost any sort of costumed history – rather than referring to a specific event that is 're-enacted'. At outdoor museums 'reenactments' are recycled daily. In history documentaries and reality history television, we regularly see re-enactors propping up claims of 'serious history'. There is an increasing amount of audience trust in re-enactors' history work, and as I have argued elsewhere, a strong desire to learn about the past in the form of a stage-play, rather than a museum exhibit label.[3]

Even though reenactments are focussed on historical accuracy and have gained increased acceptance as an accessible, popular form of commemoration, performance has not replaced monuments. In Australia, monuments have been a remarkably singular and relatively uncontested form of commemoration and public remembrance. This has largely been due to the extensive memorialisation of the First World War, where local monuments stood in as 'community graveyards'. These monuments – usually sculptural figures of the 'unknown soldier' or 'digger' – became the centrepiece of the quasi-religious annual commemorative ceremonies of ANZAC Day. (Australian and New Zealand Army Corps) They also became exalted when ANZAC histories were linked to the formation of the modern Australian nation. This combination has shaped a particular commemorative structure centred on the monument as a 'sacred space' that has helped define reenactments' status in Australia as celebratory, rather than ceremonial history.[4] Reenactments have rarely been turned to for more sombre moments of mourning. Yet spectacular military battles are eminently 're-enactable' and have long been a staple diet in reenactments in Europe and the US. Most of these battlefields, to the chagrin of many re-enactors, have been marked with monuments to the dead. This makes them unavailable as authentic stages and backdrops to an accurate recreation of events.

Figure 4.2 A monument to the 1804 Battle of Vinegar Hill

To examine this sort of reenactment in more detail, it is worth considering a 'battle' that has until relatively recently been 'unmarked' and quite unknown – the 1804 Battle of Vinegar Hill. This status, in an Australian historiography full of overseas battles and a reluctance to look to colonial conflict in its own backyard, created some interesting commemorative responses (Figure 4.2).

The 1804 Battle of Vinegar Hill was the end result of several years of attempted breakouts and rebellions by the large numbers of (mostly) Irish convicts in early colonial Sydney. Rather thoughtlessly perhaps, the British government exiled many ringleaders of the 1798 Irish Rebellion to Australia. The firebrand orator and trained soldier Philip Cunningham led around 300 convicts from the Castle Hill Government Farm on a march to 'freedom' under the revolutionary catchcry 'Give me Liberty, or give me Death!' Government forces under Major George Johnston met them near Rouse Hill just west of Sydney and gave around 20 of the rebels the latter option of their request, on the spot. Several others were executed and their bodies hung in chains for years beside the road as a rather successful warning to any future rebellions. (Although this didn't stop Major Johnston himself overthrowing poor old Governor Bligh in his second experience of mutiny after the famous *Bounty*.) The 'battle' quickly became labelled Vinegar Hill after the last battle of the failed 1798 uprising in Ireland.[5]

The bones of the convict dead that had been on the roadside also served as a warning to other convicts contemplating breaking out of the gaol that was Sydney town. Apparently they were gathered up and discretely buried by an unknown group of women many years later. Until relatively recently, Australians shunned their convict ancestry. Vinegar Hill was never regarded as an event of any historical importance. It was a sordid convict affair – a 'skirmish' with a handful of Irishmen.

Apart from the 1854 Eureka Stockade gold miners 'uprising', there are very few other moments of conflict that can be commemorated in the European Australian anti-colonial political struggle. Like the Eureka Stockade, Vinegar Hill became lost in the grand narratives of Australian history that, as noted, increasingly focused on the First World War as the cornerstone of Australia's 'birth' as a nation. ANZAC histories writ large as sacrifice for the nation upon the global stage were more suitable here than the politics of convicts or miners.

However, after a 1970s shift in understanding family history as part of national history, convicts became acceptable, even desirable, ancestors. Suddenly local historians in western Sydney became more vocal about what they knew they had on their doorstep – the site of a nationally significant story of convict struggle, and one that actually involved soldiers, rebellion and bloodshed; all the usual ingredients of the 'births' of western nations.

Local historians' claims about the significance of Vinegar Hill were heard by Irish-Australians as well as family-history researchers who found their names among the only convicts recorded at the battle; the dead and the punished. In 1984, after being hounded by local history groups, and in a climate of 'history fever' due to the impending 1988 bicentenary of the foundational moment of the 1788 First Fleet, the local Council decided to conduct a competition – with a then-substantial $5000 prize – for the design of a monument to the battle.[6]

Memorials often mark a place as well as refer to an event. Yet the exact location of the Battle of Vinegar Hill has never been confirmed. During the 1980s, at times venomous local history debates raged over the site of the battle. With scant historical information and no archaeology, after much hard investigative and conjectural work historians eventually narrowed down the likely area of the battlefield to a hillside on the Old Windsor Road. Perhaps somewhat ironically, yet also quite fortunately considering current intensive urban development in the region, the battlefield was found to be in the middle of a Lawn Cemetery. The Castlebrook Lawn Cemetery at Rouse Hill duly granted 20 square metres to erect a memorial and promised to 'maintain it in perpetuity'.

A monument to the dead of the Battle of Vinegar Hill could easily co-exist with the rose gardens, lawns and plaques of an outdoor cemetery.

Thus by 1988 a monument was in place, even if it was never completely certain that this was the true site. Almost all stakeholders were happy – except the few who were convinced the monument was in the wrong place. In front of a gathering of 200 hardy souls on a windy March afternoon, the ex-Prime Minister Gough Whitlam unveiled the monument, declaring perhaps rather grandiosely that battle of Vinegar Hill was about 'the right of all men to the religion of their choice and equality before the law'.[7]

Since the 1980s the annual ceremony has attracted an array of luminaries of the Australian political left as guest speakers including: former government minister, Al Grasby; eminent historian, Manning Clark; leading magistrate, Pat O'Shane; and Human Rights Commissioner, Marcus Enfield. It may have seemed certain during the 1990s that Vinegar Hill was securely aligned with a history of Australian radicalism. Yet there had been very little scholarly attention to the actual historical events and suggestions arose that the uprising was merely a drunken affair of looting and pillaging combined with a mass escape from prison.

Meanwhile the monument was under pressure from a different direction. It was regarded as too abstract and local historians increasingly felt it did not adequately tell its historical lesson. Between 1989 and 2000, several plaques had to be added that would better explain the monument to its viewers. Other elements such as flagpoles were added year by year so that a small 'grove' of memorial objects began to appear. At one annual commemoration, people turned up to find that uninvited guests had anonymously added their own small monument to the collection.

Despite the vicissitudes, as a place of annual remembrance ceremonies, the Vinegar Hill memorial appears to have worked well for the local history community. So too the 'Friends of Vinegar Hill' – an association of people interested in commemorating the Irish in Australia – had achieved what they had been lobbying for: a monument to the Irish dead. However the monument then quickly faded into obscurity. It sat in the lawn cemetery and became a place for small annual commemorative gatherings. Few people in Australia had ever heard of the fact that there was a major uprising in the Colony in 1804 that arguably could have overthrown the British hold in New South Wales.

For those interested in 'forgotten' histories, the drive for a monument as a marker to history appears to have a lot to do with settling scores. All the important histories, battles and struggles have monuments, so 'we must have one too'. At Vinegar Hill everything appeared to be settled;

the dead convicts had their burial marker, the job of history was done. Like the intense current interest in archaeological excavations of First World War mass graves that might put names to some of the many Australian 'unknown soldiers', the lack of a true headstone cannot be substituted by a monument.

But then there never seems to be enough in a monument to go around. Shifts in historical understanding or political interests in the past often mean more monuments need to be added to the collection. Following the excessive memorialisation of the First World War – where every town in the county had its ANZAC statue or roll of honour – we have come to expect monuments to be the solution to both grieving and righting historical wrongs. The form of the monument inspires confidence in a respectful and lasting statement. It also generates a congregation point for anniversary moments. Yet the other objective of the memorial to the Battle of Vinegar Hill was to 'raise the knowledge and profile of Vinegar Hill as part of Australian History'.[8] Arguably, this failed abysmally. Like the thousands of monuments and memorials to other local histories across the nation, the Vinegar Hill monument in the Lawn Cemetery blends seamlessly into the surrounding landscape.

Even at the time of the two-hundredth anniversary in 2004, when a remarkably sudden and surprisingly large amount of interest in the battle occurred, the commemoration ceremony at the monument attracted few people. However nearby, just two days later, thousands of people turned out on a wet Sydney day to see a reenactment of the battle (Figure 4.3).

Figure 4.3 Convicts prepare to do battle with Redcoats

In March 2004, two hundred years to the day, a reenactment of the Battle of Vinegar Hill was staged at Rouse Hill in western Sydney. Unlike the unveiling of the monument, the promise of the pageantry and spectacle of a reenactment generated significant popular interest and awareness. With the same numbers of historical participants and on a two-hundred-year anniversary moment, the staging of a battle that most people had never even heard of attracted national and international media attention.

As part of a rare breed in Sydney – being both a professional historian and a re-enactor – I was lucky enough to be involved in scripting the performance, as well as being one of the main characters, albeit with a very bad Irish accent! In a typical act of re-enactor obsessiveness, I commenced researching the event looking for evidence of not just what happened, but also how it would have *looked*.

This meant trying to visualise each and every element of the battle, the lead up and the aftermath. There is something different to usual history work about being forced to recreate the past in human scale, as performance, and under public scrutiny. You must – like our 'mobile monument' Civil War re-enactor – look at every detail, in detail. You must build a picture of an unfolding series of events in your mind's eye and then test it out on the ground. And unlike film and television or writing a history book where you can edit your work, there is the unpredictable nature of live performance.

So with a re-enactor's eye, I looked at the surprisingly little surviving historical evidence about the battle. One critical element, and generally overlooked as apocryphal, is a painting in miniature of the battle. An unattributed 'circa 1804' watercolour of the battle telescopes several moments into one image and is often labelled a 'political cartoon'.[9] However to my surprise, on close investigation I found a rich detail in the painting that actually appeared to be quite historically accurate. There were too many things correct about the minutiae of this painting to dismiss it purely as an image concocted for political propaganda. Thus with this new found assistance in visualising the battle, I decided we could do no worse than to stage a good old fashioned *tableaux vivant*; in effect to 'bring the painting to life'. Although the *tableaux* – recreating an historical image with live people – is a well worn nineteenth-century conceit that arguably says more about our ability to replicate than it might say about the historical event, especially when combined with excellent props, setting and performers, there can be some interesting results.

In their mission to achieve an authentic *mise-en-scène*, re-enactors abhor anachronisms. They, or I should say we, strive towards a complete, visible form of authenticity. In a so-called living history setting, nothing should appear out of time and place. Unlike historical pageants and parades of the past that transported history anywhere without the need for an evocative landscape, re-enacting requires credible settings for increasingly sophisticated and cynical audiences and participants.

In reenactments the combinations of props, stage, and performance expand the range of 'memes' or reassuring signifiers for history – the sights and sounds that people come to listen to and see.[10] Indeed the extent of a multi-sensory immersion for both participants and audiences is often the measure of a successful reenactment event. Reenactments, as Dennis Hall has noted, do tend to engage the senses rather than the mind, and can substitute processes of association for analysis.[11] However, as I have argued elsewhere, this process can in some ways be seen as part of a failure of other more traditional areas of history such as museums, to foster a sense of connectedness to the past for audiences hungry for history.[12]

What occurred at the Battle of Vinegar Hill was perhaps that the reenactment was actually too successful. I cringed afterwards as television journalists interviewed participants and spectators who recounted how 'that was just what it would have been like' and that 'it made me feel as if I was really there' and 'it was great to see the battle re-enacted right here on the actual site' – which was not the case at all. As the battlefield was in a cemetery, we used a local park a couple of kilometres up the road.

In many ways the reenactment of the Battle of Vinegar Hill performed a cathartic operation for this particular local history. It was forgotten history finally staged as part of the national story. There were some very concerned participants with strong emotional connections to the histories represented at Vinegar Hill. We had Irishmen playing Irish rebels, and descendants of the New South Wales Corps soldiers pitted up against them. Indeed at the height of the battle we were forced to contain the Irish from actually trying to get even, and had to remind them that this was a reenactment and that they had lost!

The Vinegar Hill reenactment was roundly held up by audiences, participants and media as a great success. It educated, entertained, and it was respectful in commemoration. The reenactment brought a spotlight to the historical event that no monument had done, or arguably could. The limitations of the monument as a commemorative form have long been noted by scholars. So-called Anti-monuments and Counter-monuments most notably appeared in 1980–90s Germany as a form of memorial space conceived to challenge the very premise of the monument. They were designed to be ephemeral rather than permanent and to deconstruct rather than displace memory. According to Noam Lupu counter-monuments 'were designed to reimpose memorial agency and active involvement on the German public'.[13] Yet these 'anti-monuments' failed to escape the symbology they resented in traditional monumental forms and thereby failed, as Lupu suggests, to create a sphere of social interaction outside the didacticism of traditional monuments.[14]

Monuments and memorials tend to be unambiguous or at least much clearer and more direct in their symbolic repertoires than reenactment or other commemorative performances of history. They often produce, shape and homogenise public memory and in doing so, facilitate forgetting as much as they evoke memory. They have been important in aligning individual responsibility and collective purpose with a material form of the national landscape.

The proliferation of monuments in Western societies during the nineteenth- and twentieth centuries has left a dense legacy of what could be described as 'funerary architecture' in urban open spaces and rural battlefield landscapes. In the US, the Gettysburg battlefield – where our 'mobile monument' was performing – has the densest concentration of monuments per acre in the world. At Gettysburg now, new dilemmas arise as these monuments grow old, contemporary historical sensibilities change, and new ways of remembering appear. Our 'mobile monument' does not make sense in such a landscape. She requires 'authentic stages' of how the site would have looked in 1863.

The custodian of the Gettysburg Battlefield, the US National Parks Service must respond to publics who want to journey through evocative historical landscapes and view historical reenactments. Parks are now busy tearing down 1970s viewing towers and recreating historical buildings, fence-lines and tree-plantings. After more than 30 years of trying to avoid it as inappropriate for solemn commemoration, they are increasingly sympathetic towards historical reenactments. Re-enacting, like historical tourism, requires a site to walk across and 'see' history both in our mind's eye and before our eyes. To retrace the footsteps of historical participants is a burgeoning pastime for cultural tourism. Watching re-enactors retracing these steps, in costume, with props, and on the same ground, is an even more attractive proposition to many.

Perhaps this urge towards a visual, participatory history is a seductive version of the past – of experiencing history rather than reading it. The seduction is directly correlated to the more distant we become from the past, the more mediated the past is for us through visual technologies, and the more it is roped off from us in museums. Yet there may be other factors at work here.

As has been increasingly noted by historians, the processes of writing history are quite analogous to the processes of reenactment. We don't question the historian who walks across a landscape to get a better 'feel' for the historical events that occurred there. We don't question the historian who writes in a prose designed to create a picture of the past in our 'mind's eye'. We expect that the historian's work will be, like a re-enactor's, authentic to the past.

It would be disingenuous to argue that monuments are worthless and should be thrown into graveyards. They have in general changed very much from their classical forms of realism and symbolism in granite and steel. Some monuments can find fresh lives – such as the Canadian Explorer with a Native American guide kneeling at his feet that graced Ottawa's public space until recently. The monument was cut in half and the guide now has his own column just a way a little.[15]

'Mobile monuments' are not some sort of complete alternative to the monument. While 'anti-monuments' have offered one form of escape from a fascination with monumentalism, reenactments offer another. As they become more attentive to an experienced authenticity, they increasingly offer a participatory, democratic form of history-making as remembrance. So too the ephemeral performances of re-enactors do not remain solid and insistent on the landscape. Physical monuments can reopen old wounds, whereas reenactments can be relocated.

Recent studies of collective remembrance have suggested that shared memory would be much impoverished if it could not reside in physically distinct spaces such as cemeteries, memorials, monuments or other less formalised spaces and sites.[16] In many ways I disagree. Shared memories of performances – perhaps best combined with preserved sites for staging – might be less divisive and less (or sometimes more), insistent.

The apparent contradiction of a person being a monument unsettles our usual understanding of the fixed nature of monuments, as well as notions of how to create a sombre and sacred commemorative space. Historical reenactment can often convey a heightened sense of ethical conflict and negotiation: the politics of individual and collective historical representations are often sharpened because reenactment is live performance. Reenactments are not so tidy and ordered as monuments. Reenactments open up possibilities that allow history to be, as is its want, unfinished business. They are also quite useful in getting to the heart of those events from the past that appear to be left outstanding in the present.

Unlike many other forms of public history, performing the past retains its marks of production but also contains the possibility of change.

Mobile monuments might just open up avenues for more inclusive and historically accurate performances that allow less didactic and more open-ended forms of commemorative ritual.

Notes

1. S. Gapps (2003) 'Performing the Past: A Cultural History of Historical Re-enactment' (Ph.D. dissertation, University of Technology, Sydney), pp. 57–8.
2. See J. Cullen (1995) *The Civil War in Popular Culture: A Reusable Past* (Washington DC: Smithsonian Institution Press); D. Hall (1994) 'Civil War Reenactors and the Postmodern Sense of History', *The Journal of American Culture*, 17:3, pp. 7–11.
3. Gapps, 'Performing the Past'; S. Gapps (2003) 'Authenticity Matters: Historical Re-enactment and Australian Attitudes to the Past', *Australian Cultural History*, 22, pp.105–16.
4. See S. K. Inglis (1998) *Sacred Places: War Memorials in the Australian Landscape* (Melbourne: Miegunyah Press).
5. See L. Ramsay-Silver (2002) *The Battle of Vinegar Hill: Australia's Irish Rebellion* (Sydney: Watermark Press) and A. Whittaker (1994) *Unfinished Revolution: United Irishmen in New South Wales 1800–1810* (Sydney: Crossing Press).
6. See the author's personal collection of Vinegar Hill ephemera 1980–2004. The vast array of monuments and memorials erected under Bicentenary funding schemes gifted Australia with hundreds of now forgotten plaques and signs – although the commemorative 'facilities' such as picnic grounds and toilet blocks are still quite useful. The sorts of events chosen were often dubious or even bizarre – such as an application for a plaque by Wollongong Council to commemorate the fact Captain Cook *nearly* landed at Wollongong.
7. *Vinegar Hill Program of Events* (4 March 1988), Vinegar Hill ephemera 1980–2004 from the author's personal collection.
8. Ibid.
9. See unknown artist (1804) [Convict Uprising at Castle Hill, 1804 also known as *The Battle of Vinegar Hill*], National Library of Australia, available at http://nla.gov.au/nla.pic-an5577479.
10. H. Schwarz (1996) *The Culture of the Copy: Striking Likenesses, Unreasonable Facsimilies* (New York: Zone Books), pp. 273–5.
11. Hall, 'Civil War Reenactors and the Postmodern Sense of History', p. 8.
12. Gapps, 'Performing the Past'; Gapps, 'Authenticity Matters'; S. Gapps (2007) 'Adventures in the Colony: Big Brother meets Survivor in Period Costume', *Film and History*, 37:1, pp. 67–72.
13. N. Lupu (2003) Memory Vanished, Absent, and Confined: The Countermemorial Project in 1980s and 1990s Germany', *History & Memory*, 15:2, p. 131.
14. Ibid., p.132.
15. An account of this event is available at http://www.urbanphoto.net/blog/2007/06/07/ottawa-a-monumental-dilemma/.
16. Irwin-Zarecka (1994) *Frames of Remembrance: The Dynamics of Collective Memory* (New Brunswick, NJ: Transaction Publishers), p. 150.

5
What Should We Do about Slavery? Slavery, Abolition and Public History

James Walvin

The two hundredth anniversary of the abolition of the slave trade in 2007 was marked by a host of British institutions. The Abolition Act of 1807 has been described as the most important Act of Parliament ever. Not surprisingly, many major state and civic institutions offered their own distinctive version of 1807. The Houses of Parliament led the way, followed, among others, by the British Museum, the Victoria and Albert Museum and the National Maritime Museum (both Greenwich and Liverpool). In fact dozens of institutions, down to tiny local schools, offered their own interpretation of the events leading to 1807 – and the significance of that Act. All this was in addition to a veritable blizzard of media coverage, and publications plus hundreds of lectures and a string of academic gatherings. Community-based organizations, ranging from the 'Equiano Society' to Wilberforce's old school in Pocklington, were equally keen to join in. Never, at any point in my adult lifetime, has slavery and the slave trade occupied so central a place in such a broadly based social and political debate as it did in Britain in 2007.[1]

Naturally enough, the discussion about 1807–2007 was fraught with complications and problems. Many groups, keen to offer their own commemoration of abolition have been taken aback to discover that slavery poses a complexity of unusual problems. But this has long been known by anyone working in the 'public history' of slavery.[2] Academic historians, drafted in to advise, are often taken aback by the contentiousness of the arguments which flare up around public discussions of slavery. Even when confronting an apparently laudable event – the ending of the Atlantic slave trade – the uninitiated are likely to be surprised by the consequent debates, by the bitterness of the arguments, and by the directions they take.

The basic point is obvious: the history of Atlantic slavery is, from beginning to end (and even beyond, into the days of post-slavery freedom), a highly contested area. Moreover here is terrain in which large numbers of groups and individuals can legitimately claim some form of stake. Academics cannot fall back on the privilege of expertise and specialized knowledge in the hope of fending off critical voices from quarters they normally do not have to engage with. If the difficulties of bringing academic expertise to bear on so contested and sensitive a public issue are troublesome, they also offer their own rewards. Think of the relatively small number of people who have traditional access to scholarly work – and compare that to the hoards passing through a museum, gallery, watching a television screen or the millions listening to the BBC World Service. Hundreds of thousands of people have trudged through the Slave Trade wing in Liverpool's Maritime Museum in the past decade. Yet, until 1992, Liverpool paid little attention to that port's remarkable entanglement with Atlantic slavery.

The anniversary thus brings into sharp relief issues of reenactment and representation of a contested and complex history in a form suitable for public consumption. Under the rubric of what has been called 'new museology', many museums and galleries have been striving to offer nuanced and multi-vocal displays, and have been responding to the increasing challenges of their place in the educational system. Students visit a host of sites from state-funded institutions to places of historical tourism as part of an educational approach that increasingly involves 'immersion', 'living history', and what is known is known as 'experiential' education.[3] In some cases, visitors and students alike are being invited to participate in 'second-person narratives'. Scripted parts are designed to increase understanding as a result of a better appreciation of context; unscripted roles prompt participants to draw on their own experience, which may or may not be in contrast to the historical record. In this way they become what Augusto Boal has called 'spect-actors'.[4] For those interested in reenactment, it is significant to note that these activities all dabble with affect, in some cases this is intentional, in others an accidental result. What they have in common is a failure to reflect on the consequences of affect for historical understanding. Here, special attention will be given to a selection of museums and sites for historical tourism looking for the effect of affect.

Context

Over the past 30 years, the subject of chattel slavery in the Americas has developed from one which seemed (in the UK at least) a marginal topic, relevant to Americanists and Africanists, rather than European historians. Today it is impossible to keep abreast of the relevant literature. Major conferences devoted to slavery pop up in all corners of the globe. On the back of Atlantic slavery, there has been a revitalization of slave studies from the classical world to the twentieth century.[5] It is an indication of the central position now occupied by black slavery that, in both the popular and educated mind, when we speak of slavery it is widely assumed we mean *black* slavery. Slaves were black: to be black was to be enslaved. Of course this was not true in a host of slave societies. It is also an indication of the powerful role of black slavery in the public imagination that it is commonly assumed that the ending of black slavery in the Americas (finally in Brazil as late as 1888) brought slavery to an end.[6] Yet *Anti-Slavery International*, the modern-day descendant of the Anti-Slavery Society founded in 1829, continues to thrive in London, overwhelmed by the task of campaigning against slave systems the world over. Slavery lives on.

Nonetheless the Atlantic slave system has captured the imagination, not least because few areas of modern history can match the litany of human suffering which is the core of Atlantic slavery – especially of the Atlantic slave trade itself. But placed in the long history of slavery, the highly racialized slave systems of the Atlantic were perhaps the *exception* rather than the rule. Their chronological proximity and their pervasive consequences down to the present day explain why slavery is popularly equated with Africans. The demographic data also helps to explain how ethnicity and slavery have become so confused. About 12 million Africans were loaded onto the Atlantic slave ships, and about 10.5 million Africans survived the Atlantic crossing, spread over a period of almost four centuries. We cannot tell how many enslaved people died in Africa itself en route to the coast. The statistics can be sliced in any number of ways. By 1820 for example some 11 million people had crossed the Atlantic to settle in the Americas. But of these, only about two and a half million and were European: the rest were African.[7] Across huge swathes of the Americas, especially in Brazil and the Caribbean, it was the African, not the European, who was the key pioneer of settlement. And all this for the economic betterment of European settlers and their political and economic sponsors.

It would be wrong to imagine that these events – played out largely on the African coast, in the Atlantic and in the Americas – had only a marginal role in British history. Until recently however there was a tendency to consider this story of Atlantic slavery as a distant, foreign episode; out of sight and generally out of mind. In part this was a consequence of academic specialisms, and the understandable sense that slavery was an area best left to Africanists, Americanists or maritime historians. What has emerged in recent years is a growing awareness of the centrality of slavery to the British historical experience. It is now accepted that what happened in the enslaved Atlantic was integral to the emergence of Britain itself in the period, say 1655–1807.

That basic point is most obvious when we consider the history of the major slave trading ports. By turns, London, Bristol and Liverpool dominated that trade at different periods, but there were a host of smaller ports involved, some of them, today, unlikely candidates: who would imagine Poole or Lyme Regis as slaving ports? We now know that of the 27,000 slave trading voyages, about 11,000 were British (or British colonial) and of those almost 6000 originated from Liverpool.

The holds of those outward-bound ships, destined for trade and barter along the African coast, were packed with goods drawn largely from the economic hinterland of the home port. But they also carried manufactured goods and produce from throughout Britain, alongside items transhipped from Europe (French wines) and Asia (Indian textiles, and cowry shells from the Maldives). As the trade with Africa grew, and as slavery yielded a growing bounty to the adventurous (and the lucky), involvement in the Atlantic trade proved irresistible for ever more people in all corners of British life.[8] Those who resisted the lure of slave trading – notably the Quakers – seemed odd in suggesting that ethical and religious issues should override profits. Few sided with the Quakers, initially at least, preferring instead to accept the assumption that slavery was the source of very great well-being to the British. Indeed the debates inside Parliament, from 1690 onwards, were primarily how best to regulate and encourage the trade in African humanity.[9]

Slaves and slavery also flitted in and out of British life. The evidence is there for all to see. The legal complexities of slaves in England periodically taxed English courts: black faces can readily be found in seventeenth- and eighteenth-century portraiture, and Africans and their descendants appear in parish registers, and on graveyard headstones.[10] More impressive, though sometimes not obvious, the rewards from slave labour can be seen in a range of British buildings, few more spectacular than Harewood House in Yorkshire (Lord Harewood's ancestors made their millions in the sugar trade). Less noticed perhaps was the impact of slave-grown produce on European social life. Slaves hovered, ghost-like, over British social life in those rituals of sweet-tea drinking, and the masculine world of tobacco culture.[11]

The British Atlantic slave system was kept intact by British military and maritime power. The Royal Naval presence in the Atlantic and along the Caribbean Sea lanes secured the vital flow of supplies of Africans. The Navy also ensured that the Caribbean slave colonies remained armed against the permanently rebellious instincts of their enslaved populations. It was no accident that all the great British naval heroes of the eighteenth century had learned their craft in Caribbean waters. The Africans were of course ferried into the Americas on a massive flotilla of private, commercial merchant ships. And by the mid-eighteenth century this trade was dominated by Liverpool. But it would have been impossible to appreciate that fact by visiting the Liverpool Maritime Museum before 1992.

Liverpool

Liverpool is home to an old black community, with strong nineteenth-century links both to West Africa and the Caribbean. But initially there was scant mention of this, or of the slave trade in the Maritime Museum. Perched above a sack of sugar, in an obscure corner of the Museum, there was a map showing a simple triangular Atlantic trade. And that was it. This failure to engage with Liverpool's slave trading past was made good by the determination of one of Liverpool's most wealthy men – Peter Moores. Moores had inherited from his father the Littlewoods business empire which included football pools and a chain of department stores; he was also patron of various philanthropic foundations. Convinced that slavery had been 'a taboo subject' in Liverpool, Moores put up a large sum to establish a slave trade wing within the existing Liverpool Maritime Museum. I was invited to join a team of guest curators on a project which was to prove difficult, troublesome and sometimes worrying. It was soon apparent that writing a book about slavery was simplicity itself compared to mounting a public exhibition about it.

The problems were manifold (and obvious) as soon as you begin to think about such an exhibition. One local Professor of History (a distinguished Africanist) cautioned us not to try: we should decline the offer and presumably let England's premier slave port continue to ignore its eighteenth-century history. (By 1807 for example, one African in five crossed the Atlantic in a Liverpool ship.) With a large sum to spend, the museum and its advisors were not about to take that route. But the topic – slavery and Liverpool – forms a brew of just about every contentious issue you might wish to discuss: race, guilt, compensation, city politics, schooling, community relations, urban deprivation, unemployment, and gender. And all this in addition to whatever arguments the transient academics might want to air.

The benefactor had a small specialized committee liaising with the museum (itself responsible to the broader national maritime museum network.) The museum appointed a group of historians to advise. Initially that group was overwhelmingly white and predominantly male. The decision to incorporate more women, more blacks, raised further questions: should we turn to African men and women? One black committee member was African, but had lived in the US for many years and his American accent displeased some members of the local community, with whom we had periodic discussions. For some reason or other, accent also mattered.

There we faced a hostile and sometimes an aggressive reception ranging from root-and-branch opposition to the whole idea, to a demand that the museum should be moved to the black community in Toxteth (that is, up a hill), through to more manageable points that training and employment ought to be offered to members of the local black community. However the existence of a predominantly white advisory group, talking about black history, remained a permanent irritant. Arguments about 'insiders and outsiders' surfaced regularly: about the rights and wrongs of outsiders (outsiders by ethnicity, regional or national origins – and sensibility) versus insiders. It was an argument familiar to anyone who had witnessed the history of the US in the 1960s and 1970s. But for those of us, like myself, who were outsiders, it was an unwinnable argument: intellectually crippling, historically restrictive and yet not completely pointless.

No less time-consuming were the debates among the historians about how to represent the interior of the slave ships. How could we characterize or re-enact human violations on that scale: should we even try to convey the stable-like squalor of a mid-oceanic slave ship, the living and the dead chained together, pitching and rolling in their own filth for weeks on end? It quickly emerged that historians are not very good at dealing with such issues. The abstract analysis of the numbers of Africans involved, (the death and sick rates, survival rates, timings of crossings, all and more) seem morally neutral when stripped of their humanity. Efforts to present accounts of the stink of a slave ship, the shrieks of the mad and the distressed, and the agonies of the dying raised the fundamental question of our obligation to the visiting public. Do we want to rub the visitors' noses in the slave mire in order to make a (perfectly valid) point? What role should affect play in historical sites? These issues are central to the debate among historians about the uses of historical reenactment. Some museums, galleries and tourist sites have experimented with the use of sensory experience such as sound and smell and other forms of 'immersion'. Notwithstanding the broader issues of historical accuracy – how do we know what the past smelt like, and how do we know what historical actors considered to be a bad smell? As Norbert Elias has famously shown, the 'threshold of repugnance' has shifted dramatically over time.[12] Are there others ways of addressing the issue?

In the event, an easier, less sensational route was chosen: of flickering images in a darkened room to the sound of a creaking ship and hushed voices. Even in so muted a form, this proved to be one section of the exhibit which often caused the greatest distress to black visitors. It stands as a reminder that poignant effectiveness often flows from gently made arguments.[13] Re-enacting the past then, even within the safe walls of the museum, can provoke a profound emotional response. The question remains however, does it increase historical understanding?

The outcome

The slave trade gallery doubled the numbers of people visiting the Liverpool Maritime Museum in its first twelve months (1994–5). Numbers have, inevitably trailed off, but continue at a healthy and responsive level. The notice board, inviting visitors' written comments (all of which were kept and analyzed) has proved remarkably revealing. My own worries that they might attract racist remarks were ill-founded. Among school visitors, reactions have been overwhelmingly positive. Overall the slave trade wing has proved a great success, with favourable international publicity, and a steady stream of VIPs. In 2007 this gallery made way for a new museum of slavery – and all on the back of 18/07/2007 and the status of European City of Culture for Liverpool in 2008.

Pulling punches

For a decade, here was an exhibition which, of necessity, had pulled its punches. How could it be otherwise? For a start it represents work by a committee, monitored by a host museum, itself responsible to a benefactor who had his own vision of what was required. The flurry of substantial memos and briefing papers which all the academic advisors were expected to produce (some of which ran to thesis-like length) were reduced, in the end, to simple captions. The art of museum caption-writing is more akin to drafting a newspaper headline than developing a scholarly argument. Moreover, academics can, within the limits of our own self-determined argument, say what we want. We need not worry about whether people like or approve of what we say or write; whether it will ruffle sections of an urban community. We don't concern ourselves, normally, with how the casual visitors might react to our ideas (not least because they rarely encounter them). Providing we accord to the conventions of scholarship, we can write and say more or less what we want. I doubt that this is possible in the more exposed and communally responsive environment of public history.

The problems of public history

Though the difficulties of discussing the history of slavery are similar to those encountered in other fields of historical reconstruction, slavery poses its own distinct and peculiar difficulties. Today there is an odd juxtaposition of slavery and modern tourism. Slavery thrived in many regions which are, today, attractive tourist destinations. Parts of the US, and especially the Caribbean, try to utilize their past to add to their commercial/tourist attractions. But how do you talk about slavery in terms which are accurate, decent and yet not rebarbative? The Caribbean poses special difficulties for anyone keen to combine a sense of the past with natural tourist attractions, for here was the centre of some of the worst slave violations. Discussing the slave experience with bus and boat loads of tourists leads to an inevitable sanitizing of the story, or even to overlooking aspects of the past which are too uncomfortable to present to tourists. This is particularly striking on plantations.

The plantation was the crucible in which raw African muscle was harnessed by a brutal regime to the task of tapping the economic potential of the region, and all for the economic betterment of white settlers and their European backers. It is clear enough that, notwithstanding variations (determined largely by the nature of the local crop), the plantation represented the slaves' time on the cross. Even in the more benign culture of tobacco (as opposed to sugar for example), the levels of brutality doled out to the slaves were astonishing – even by eighteenth-century standards. Plantations were developed in fertile tropical and sub-tropical regions, at a number of sites which today are attractive destinations for tourists in search of natural beauty, sunshine and a taste of the romantic past. Yet there is nothing romantic about black chattel slavery.

Consequently the problem of historical reconstruction on old plantations has often been resolved by dissembling: not so much distorting, but forgetting.

It normally begins with the brochure handed out as you buy your entry ticket. For more than 20 years I have collected brochures and advertisements from tourists sites housed on former slave sites, and the deceptions are commonplace. Let me give you some examples. The Belvedere Estate in Jamaica, 'invites you to take a step back in time' (but not *too* far back of course in case you trip over a slave). Another telling phrase invites us to visit 'our historical ruins', which turn out to be the factory and the Great House, not the slave cabins. A picture of a woman dressed in African slave dress, pounding ingredients appears under the heading, 'Consult our herbalist'. She would almost certainly have been an *obeah* women (an African medicine woman), the very person planters feared, and persecuted because of the mysterious power she exercised over the slaves. Yet today, she is offered as a quaint folk figure, neutralized of her historical importance.[14] Herein lays the potential and the danger of reenactment as an activity purporting to offer an educational rather than an entertaining experience.

Time and again, the antiquity of a plantation is represented by its fine old buildings, or in dynastic terms. Or the Great House has been restored to its former glory. Or the furnishings of the Great House are `traditional planters' furnishings'. At Good Hope estate, some 15 miles from Montego Bay on the north coast of Jamaica, a group of wealthy investors have converted the property into an expensive hotel for prosperous (mainly North American and Japanese) tourists. Here the emphasis is on location, peacefulness and furnishings. They stress the property's wonderful outlook, which is indeed stunning. But anyone with a historical sense will immediately grasp another major point. This late-eighteenth-century Great House, lavishly constructed from dressed stone by an army of skilled workers, was actually built by slaves – as were the furnishings. And the whole enterprise of course was made possible by the fruits of slave labour, in the fields which fall away from the Great House down to the river: thence to the sugar markets of Britain. But there is another, less obvious story, lurking on the horizon. If you look to the south, you will see the impenetrable range of interior mountains and jungle – 'the cockpit country' – which provided an escape for runaway slaves, and which came to house runaway communities. Those communities pressed hard on slave properties like Good Hope. For all their wealth and domestic splendour, successful planters lived a precarious life, worried about their own slaves, fearful of runaways and always conscious of the revenge they might exact at any moment. Yet the tourist will only learn of this from other sources. Certainly the tourist-conscious owners don't want to let them into the secret.

Similar stories could be recited from one former slave colony to another. Even when the slave past *is* confronted, it is often skipped over lightly. In Barbados, tourists are given a booklet, *Heritage Sites of Barbados,* which breezily describes the island's history thus:

> Some ten Africans came to Barbados in 1627 as the first slaves and with the coming of the 'Sugar Revolution' in 1643–50 their numbers increased until they outnumbered the English settlers by 12 to 1. For some 207 years, Barbados was a slave plantation society.[15]

In fact, in that time, perhaps one-third of a million Africans had been landed in Barbados – an island the size of the Isle of Wight.

If you really want to see cosmetic historical surgery at its most extreme, you might visit some of the beautiful former slave properties in North America. Many promote themselves as tourist attractions via the splendours of the buildings (normally the planter's Great House), their lavish gardens, or simply through that mythical 'moonlight and magnolia' which entered popular culture mainly through the movies. For example Kent House' was built, the visitor is disarmingly informed, 'when interest rates were only 1.5 per cent'.

Or you might want to visit Middleton Place near Charleston, South Carolina, to look at 'America's oldest landscaped gardens'. The only mention of violence here is the burning of the property in the Civil War. Across the South – Old South and Deep South – the present-day owners of former slave sites emphasize the architecture, the fittings and the landscape, but rarely mention the black labour force which made everything possible. More than that, when you look at the brochures and booklets produced for the tourists, you will rarely see a black face. Normally, the frames are occupied by well-fed white people, (their beaming smiles an advertisement for costly dental care) and often dressed in 'traditional' costume. Again the potential of reenactment to obfuscate rather than inform is evident.

If you drive along the road which clings to the James River in Virginia, running east from Richmond towards the Chesapeake Bay at Norfolk, you pass through the heart of the early tobacco culture: the region which was the economic bedrock of the Old South and which yielded such prosperity to settlers and Europeans – and such miseries to armies of Africans and their local-born descendants. By 1800 there were almost one-third of a million slaves in the region. Today it is a popular tourist destination, and the river road is dotted with plantations beckoning the tourist from all angles. Visit those properties, wander through the gardens and houses, look out across the James River (the final leg of that interminable slave voyage from Africa, and the export point for the slave-grown tobacco departing for Glasgow) and you will barely find a *mention* of slaves. I have collected a fistful of literature from those properties and, from first to last, slaves are conspicuous by their absence. Again, the accompanying pictures rarely, if ever, display a black face. The plantation literature offers touching dynastic stories: loving spouses dead before their time, tales of Civil War heroics (amputations on the kitchen table), the centuries' long struggle to keep the property in good shape. But you look in vain for the people who made everything possible – the slaves.[16]

The tourist heart of the region is of course Colonial Williamsburg. And here, again, the issue of slavery is fundamental to local history. For years Colonial Williamsburg has wrestled with the problem of how best to present slavery to the tourists. The starting point must be demographic. Colonial Williamsburg has been reconstructed as it was in 1774. But in 1774, 49 per cent of the local population was black, though no tourist today would grasp that fact, however long you stay in the town. Once a marginal theme in the town's representation of itself, slavery has been shifted to a much more central location, with slave quarters, African-American churches and interpreters, and every interpreter well-briefed about the broader issue of slavery in Williamsburg and the nation at large. Williamsburg, along with other similar tourist sites in the same region (notably Monticello and Mount Vernon), has given long and costly thought to the question of slavery and to the local black presence. As Scott Magelssen has shown, a 1994 reenactment of a slave auction in Williamsburg proved to be confronting and distressful for audience and participant alike.[17] The debate how best to present the slave presence continues to pose problems (especially in the context of a continuing decline in tourist numbers.)

At both Mount Vernon and Monticello, slavery has similarly shifted from the margins to the centre of local representations. On my first visit to Monticello as a tourist in 1979, the female guide (known at the time as a 'hostess') made no mention of slavery whatsoever; even the most innocuous questions about slavery were treated as a form of intrusive vulgarity. Today, the issue of slavery is normally the *first* issue raised by the guide. There, and at Williamsburg, the local African-American guides and interpreters provide excellent accounts of slave life in both places.[18]

In the wake of the Sally Hemmings affair, it would have been impossible for Monticello to persist with its earlier marginalization of slavery.[19] But in all these three major Virginian slave sites, two of them homes of Presidents, slavery had been central to economic and social existence. Over the past generation, all three inevitably came under pressure from local and national African-American community and interest groups, and all have clearly been influenced by the ebb and flow of academic debate about slavery. Williamsburg has its own coterie of distinguished historians, and has links with the College historians on its doorstep, in addition to inviting outside scholars to engage in the debate about the representation of local slavery. Monticello is of course intimately linked to Jefferson's own foundation, the University of Virginia, while Mount Vernon nestles close to the nation's political capital. Each, for different though related reasons, has registered the seismic shifts which have transformed our understanding of the slave past over the past generation.

Today there are myriads of institutions across the US which seek in some way to memorialize slavery. The process however has been slower and more muted in the UK – for some perfectly good (and some rather bad) reasons. But the gradual movement of the history of slavery towards the centre of British cultural considerations in 2007 created an irresistible momentum. Over the course of 2007, spurred by keen governmental interest (the Deputy Prime Minister chaired a committee which oversaw the events organized for that year), the proliferation of slave trade-based events and memorabilia was extraordinary. There were special postage stamps, a commemorative 'abolition' £2 coin from the Royal Mint and – most lasting of all perhaps – a decision to include the history of the slave trade and slavery in the national curriculum for schools. The interface between the teaching of history in the classroom and its presentation in museums and sites of historical tourism is something that will be important to monitor. On top of this, the BBC (radio and television) provided massive coverage of abolition: most departments were keen to involve themselves – including comedy, religion, drama, current affairs and news, the World Service, the BBC website – all and more had their say. This represented a massive shift away from an older outlook which tended to consider slavery as distant and remote, a marginal theme in British historical interests. This shift was most noticeable when the Archbishop of Canterbury, in a sermon preached before the Queen in Westminster Abbey, listed the benefits which flowed to Britain on the back of Atlantic slavery.[20] Rowan Williams was speaking in the vein and tone of his namesake Eric.

Of course these links – these material benefits – have long been obvious – if you scratch away at the physical fabric of British life. What initially seems an unlikely object often yields enslaved foundations. Take the example of Harewood House in Yorkshire. One of the nation's most beautiful stately homes, a staggering Palladian pile, set in grounds designed by Capability Brown, its furnishings by Chippendale, and housing one of the country's finest private art collections, it is the family home of Lord Harewood, first cousin to the Queen. The Harewood Trust which runs the house is currently considering how best to use the family's history to maintain and enhance the 300,000 plus visitors needed each year to pay the bills. In origin Harewood House was built on the sugar trade, and therefore on the labour of enslaved Africans. The Harewoods transformed themselves from humble Yorkshire gentry stock (the Lascelles) into aristocrats of fabulous wealth in the period 1690–807. This was achieved via trade to and from the Caribbean, on government contracts (mainly with the Royal Navy and Customs-collecting in Barbados), on money-lending to planters, and eventually as planters (and therefore slave owners) on a staggering scale. In 1787 they owned 18 plantations on four different islands: the Harewoods sold their last Barbadian plantation in 1970.[21] But who, today, could walk through Harewood House and even *imagine* that everything hinged on the labour of Africans in the Caribbean? Who would think that Harewood House has *anything* to do with slavery? Yet to confront that fact and to attract tourists poses substantial difficulties (not least the risk of incurring the animosity of the West Indian community, ten miles down the road in Leeds.) Yet to ignore the defining fact of slavery would invite even more trouble.

Other British institutions face similar and related difficulties. The Maritime Museum in Greenwich has a new 'slavery collection', bought at substantial expense from a private collector. It contains a number of fine late eighteenth-century cartoons and caricatures, which, though familiar to students of the period, are often shockingly racist and deeply disturbing to the casual visitor: the Museum is perched on the edge of a large black neighbourhood. At a private viewing when the collection was first acquired, there was vociferous indignation from local black community leaders. The idea of displaying such material in public exhibitions seemed merely to promote continuing racist imagery. Museum staff had to give serious thought how best to present their new acquisitions, a problem unlikely to trouble an academic author who wanted to publish or write about those same images.

Harewood House and the Maritime Museum today, like the Liverpool Maritime Museum a decade earlier, are caught in the same intellectual – and political – problem: how best to deal with a topic of enormous social sensitivity. The same difficulties arose when Parliament itself decided to have an 1807–2007 exhibition in Westminster Hall. It was natural for Parliament to enter the commemorative field; after all it was an Act of Parliament which banned the slave trade after 1807. The problems were, again, complex and often unpredictable and the structure of managerial command labyrinthine (involving both Houses, both Speakers, the Parliamentary Estate and even the Monarchy – with various government departments invited to keep a watching brief on the unfolding exhibition). Any exhibition in Westminster Hall faces the daunting presence of the Hall itself, with its overpowering physical presence, and location of some of the nation's most critical historic moments.

Yet 1807 was problematic even for Parliament. How could Parliament represent the story of its abolitionist role in 1807 *without* mentioning the role of Parliament in the previous century – when it legislated *in favour* of the trade? It was a sign of how the historical study of slavery had changed in recent years that the Parliamentary Exhibition of 2007 faced a different set of intellectual considerations from those facing Liverpool 15 years earlier. First of all the slave trade itself looks very different (thanks in large part to the remarkable researches of David Eltis, David Richardson and their colleagues). We know so much more about the minutiae of the slave trade, which now seems bigger, more pervasive, and more central than we had thought previously. *And* it was at its height when the British decided to turn their back on it. Secondly abolition also looks very different. Recent studies of abolition have revealed it to be more broadly based, more popular, and to have included female and African voices in ways rarely considered earlier. Thus the story of the abolition of the slave trade has to be presented as a very different historical phenomenon than we imagined only a generation ago.[22]

These historical changes were at the heart of arguments which ranged back and forth across the country's museums, galleries, libraries and public forums in 2007, and for at least two years before. (To my knowledge only one major institution – The National Archives – decided NOT to commemorate 1807, for reasons which have never been publicly explained.) What unfolded in the course of 2007 was in effect a prolonged, national debate about abolition – but about other more profound issues as well. The year 2007 saw a public engagement about critical aspects of British history, the likes of which I had never experienced in 40 years as a professional historian. (The French had similar discussions in 1989 during their own bicentenary.) Of course the long-term effects of these commemorations are more uncertain, but there is no doubt, already, that 2007 provided an important catalyst for a remarkable public debate about history, about the nature of British involvement with the wider world, and even with the very idea of what we mean by the creation of British identity (that favourite word).

What had seemed, at first sight, a simple, unexceptional proposal – to memorialize the abolition of the slave trade in a fashion which is appropriate to specific institutions – was promptly transformed into a confused but productive cultural debate which had social and political ramifications few could predict. And the debate rumbles on: what should we do about slavery?

Notes

1. As a personal example, between January and August 2007, I delivered 60 invited lectures on abolition, in addition to dozens of radio and television interviews.
2. See for example, E. Kowaleski Wallace (2006) *The British Slave Trade and Public Memory* (New York: Columbia University Press).
3. See K. C. Barton and L. Levstik (2004) *Teaching History for the Common Good* (London: Lawrence Erlbaum Associates).
4. See A. Boal (2002) *Games for Actors and Non-Actors* (New York: Routledge).
5. See the annual bibliography on slavery published in the journal *Slavery and Abolition*.
6. For the most recent study, see D. B. Davis (2006) *Inhuman Bondage: The Rise and Fall of Slavery in the New World* (New York: Oxford University Press).
7. The statistics of the trade are to be found in D. Eltis, S. D. Behrendt, D. Richardson and H. S. Klein (2000) *The Trans-Atlantic Slave Trade: A Database on CD-Rom* (Cambridge: Cambridge University Press).
8. See K. Morgan (2001) *Slavery, Atlantic Trade and the British Economy, 1660–1800* (Cambridge; New York: Cambridge University Press).
9. See W. A. Pettigrew (2007) 'Parliament and Escalation of the Slave Trade, 1690–1717' in S. Farrell, M. Unwin and J. Walvin (eds) *The British Slave Trade: Abolition, Parliament and the People* (Edinburgh: Edinburgh University Press).
10. See D. Dabydeen, J. Gilmore and C. Jones (eds) (2007) *The Oxford Companion to British Black History* (Oxford: Oxford University Press).
11. See J. Walvin (1997) *Fruits of Empire: Exotic Produce and British Taste, 1660–1800* (Basingstoke: Macmillan Press).
12. See N. Elias (2000) *The Civilizing Process: Sociogenetic and Psychogenetic Investigations* (Oxford: Oxford University Press).
13. The publication which emerged from this enterprise is A. Tibbles (ed.) (1994) *Transatlantic Slavery: Against Human Dignity* (London: HMSO).
14. See Touren's Negril & Island Tours – Transportaion – Transfers, available at http://www.geocities.com/tourenltd/rosehall.html.

15. *Heritage Sites of Barbados* [tourist pamphlet].
16. For the other side of the story of slavery in this region, see R. Issac (2004) *Landon Carter's Uneasy Kingdom. Revolution and Rebellion on a Virginian Plantation* (Oxford: Oxford University Press).
17. See S. Magelssen (2006) 'Making History in the Second Person: Post-Touristic Considerations for Living Historical Interpretation', *Theatre Journal*, 58, pp. 295–6.
18. In fact I found them the best of all the various local guides.
19. For an account of Thomas Jefferson's alleged relationship with the slave, Sally Hemmings, see A. Gordon-Reed (1997) *Thomas Jefferson and Sally Hemings: An American Controversy* (Charlottesville: University Press of Virginia).
20. Dr Williams gave a string of lectures and sermons in March 2007, from Zanzibar to Hull, on the issue of the Church, slavery and reparations. See *Guardian* (26 March 2007); BBC News website (27 March 2007); and The Wilberforce Lecture, Hull (24 April 2007). See also A. Tyrrell and J. Walvin (2004) 'Whose History is It? Memorialising Britain's Involvement in Slavery' in P. A. Pickering and A. Tyrrell (eds) *Contested Sites: Commemoration, Memorial and Popular Politics in Nineteenth-Century Britain* (Aldershot: Ashgate), pp. 147–69.
21. J. Walvin (2005) 'The Colonial Origins of English Wealth: The Harewoods of Yorkshire', *The Journal of Caribbean History*, 39:1, pp. 38–53.
22. See the 'Illustrated Exhibition Catalogue' in Farrell, Unwin and Walvin (eds) *The British Slave Trade: Abolition, Parliament and People*.

6
Reenactment and Neo-Realism

John Brewer

What are we to do about reenactment? Here's a term that seems to cover a multitude of sins and a myriad of forms – the Christian sacrament of communion, the activities of societies for creative anachronism, Shakespeare's history plays, movies about the Alamo, art forgeries, a lot of pornography, most scientific experiments. Perhaps it is better to ask why supposedly sane academics have come to be interested in or pre-occupied by reenactment. One easy answer is to say of reenactment, as of sexually transmitted disease, that there is a lot more of it about nowadays. But reenactment has been around for 200 years or so. Its forms and frequency may have fluctuated but it has been a general feature of the culture of modernity, with its progressive view of history which figures change as both progress and loss. (Think for example of nineteenth and early twentieth-century world fairs, almost all of which contained not just evidences of modernity, but reenactments of the savage and the primitive).[1]

In fact recent academic concern with reenactment is rather more specific. It seems to me to be part of an anxiety about the proliferating interest in the past, from which its natural custodians – professional historians – have largely been excluded. (OK we all know the names of those who haven't, but they are not many more than the fingers on one hand.) Amateur enthusiasts, the representatives of identity politics and, above all, participants in the culture industries that produce TV, video, and film – not academics – are the creators, purveyors and consumers of this past. What they create is a past that is at once consumer good and cultural possession.

The responses of academics to this phenomenon – at least of those who have not dismissed reenactment as an illusory and unimportant path to historical understanding – have been threefold. The first two are anthropological in character. The post-Durkheimian world of social and cultural history seeks to analyze the functions, purpose and meaning of reenactment, and to locate it within the categories of collective identity and memory and, in its more sophisticated versions, within the realm of social conflict. Pierre Nora's series of volumes *Les Lieux de Memoire* stand as a (distinguished) representative of this trend.[2] This is a way of re-appropriating reenactment and bringing it back into the realm of scholarly history.

The second response is ethnographic or (to use Garfinkel's term) ethno-methodological (if you can't beat them, join them): to act not just as an observer of reenactment but as a (possibly privileged) participant, the fate of the historians (and other academics) who took part in *The Ship*.[3] The third response is to treat reenactment more formally, to attend to its mechanics or poetics, as in Jonathan Lamb's four-fold classification of reenactment as pageantry, theatre, house and realism.[4]

In this essay I want to pursue a path closer to Lamb's than any other – one that attends to the poetics of reenactment, though what I want to argue is that we have become preoccupied with a particular notion (a sentimental and naïvely somatic) idea of how reenactment does and should work, which has occluded other ways in which we might want both to practice and analyse reenactment. Specifically I assume that all forms of reenactment either tacitly or explicitly assume a notion of the real, and that this is the question to which we should attend. I make my argument using the case of Italian neo-realism or, more specifically, using the example of one neo-realist film, Roberto Rossellini's 1946 masterpiece, *Paisa*, released in the Anglophone world as *Paisan*. (I restrict myself to this case study in part because I don't want to get into the vexed debate, that has run since the 1940s, of what neo-realism is or was.)[5] Rather I focus on the techniques and assumptions that underpin *Paisa*'s notion of the real. My aim is not to argue that we should revert to neo-realism, which like sentimental reenactment has a very specific historical context, but to point out that any serious discussion of reenactment must address the question of the poetics of the real.

I want to begin by looking at reenactment from the point of view of the re-enactors, and then move to those who create reenactments but may not be participants in them. I realize that a distinction between participants and impresarios is not always a clear one, but it has its heuristic uses, not least in discriminating those who often wish to overlook, and those who necessarily have to address, the poetics of reenactment.

One of the fundamental, and certainly one of the most enduring historical problems has been about how to deal with the distance between the past and the present, the distinction between subject and object, participant and narrator/recorder/witness. History – as text, image or performance – constitutes the means by which these binaries are connected. My sense is that enthusiastic re-enactors embrace reenactment in part because they believe that this tricky problem can be short-circuited or avoided through the process of reenactment. Thus reenactment is advertised at one website as follows: 'Have you often wondered what it might have been like to actually live in the past? Historical re-enacting gives you that chance'.[6] Self-descriptions of re-enactors seem to collapse the distance between past and present, reducing it to zero so that the re-enactor inhabits a sort of overwhelming timelessness in which the present self and past other merge into a single identity, a unique individual experience. (As one would expect with such a sentimental view, what is important is not the truth of the enactment but its psychological effects.) Paradoxically historically specific paraphernalia are usually what make this possible – certain foods, clothes, locations – enabling the re-enactor not just to wear dead men's shoes but to inhabit their skins.[7] This is, of course, a form of fetishism, in which authentic objects supposedly provide the subject with a complete experience of the past.

On the whole re-enactors seem to see reenactment as experiential and somatic. Re-enactors typically fight battles, rehearse rituals etc.[8] They are about doing things, not about thinking things. (I don't know of too many examples of them engaging for instance in philosophical or theological arguments.) This seems to be part of a widespread assumption that language and thought are culturally and temporally specific while feeling and somatic experience are in some sense timeless, an adjunct of human nature. To fight as a seventeenth-century soldier, to sail as an eighteenth-century mariner, to cook as a nineteenth-century housewife (apologies for the sexual politics, but they are 'in period'), to suffer as a twentieth-century victim of persecution is to replicate and indeed inhabit their 'experience' through performance. Again, the issue of distance is occluded. But, as Martin Jay reminds us in his admirable recent book, *Songs of Experience*, our notion of experience also works against this sort of identification: 'the word experience has often been used to gesture toward precisely that which exceeds concepts and even language itself. It is frequently used as a marker of what is so ineffable and individual (or specific to a particular group) that it cannot be rendered in conventionally communicative terms to those who lack it.

Although we may try to share or represent what we experience, the argument goes, only the subject really knows what he or she has experienced'.[9] Re-enactors, it seems, are engaged in seeking the triumph of hope over experience. Put less factitiously, their notion of experience is an instance of what Karel Kosik has called 'the pseudo-concrete', a view that seems to have forgotten the human mind.[10]

Of course the picture produced by those whose task it is to create reenactments – I'm thinking here chiefly of those who produce historical media – TV and the like – have a somewhat more complex and sophisticated view. The creators of a commodified version of the past have to think about what strategies will best connect the product to the consumer, a process that, given the ways in which marketing strategies work, is largely seen as a question of establishing forms of identification. (I want to ask what other sorts of tactics are there now. Identity seems to have swallowed everything.)

Reenactments of this sort tend to come in two rather different flavours. The first falls into the realm of traditional political and military history (though there is also a social history version of this) and tells a story in which the audience is invited to take sides with the forces of light in their struggle with the forces of darkness. The temporal and narrative structure of such recreations is usually heroic and progressive, supposing a fairly standard idea of linear time. They often entail an unseen master narrator – think of Sir Lawrence Olivier in the 26-program ITV series *The World at War*. The attraction of such re-creations to an audience is through an identification with the goodies, a group whose character – national, ethnic or whatever – is assumed to be that of most of the audience. The nature of the identification is not necessarily sympathetic, though it often is. But the chief assumption is not that we wish to inhabit this reenactment, but that we recognize that it in some ways made or makes us what we are. Identity is more important than sympathy. Such reenactment enables us (whoever the 'us' are) to place ourselves in a larger history, to see ourselves, ordinary as we may be, as participants in or the beneficiaries of a transformation – the growth of democracy, the emancipation of minorities, the defeat of totalitarianism, the emancipation of the self and so on. As Taylor Downing, the managing director of Flashback Television puts it, talking about his mailbag, 'there are millions of intelligent and thinking people who are genuinely interested in how the past has helped to make us what we are'.[11] These large-scale narratives, rare cases in which abstract notions are purveyed to a public, are normally scaled down by concentrating on a small group of heroic figures. It is possible, however, for there to be a considerable distance between what is re-enacted and the audience. The crucial question is whether there is a linear narrative (often what post-structuralists would call a myth of origins) that can connect the past to the present.

This sort of reenactment approximates to what I, using a term derived from Jay Appleton's analysis of landscape, have called 'prospect' history.[12] What do I mean by this? Prospect history presents a single, superior point of view – a bird's eye perspective or a view from a lofty peak – in which an extensive, large-scale landscape is surveyed. The viewer or narrator is not in the picture but outside it. Subject and object are clearly differentiated and distinct. Because of height, size and distance, what is observed and recorded is general not specific, an undifferentiated shape or aggregated trend whose contours and surface can be seen, even when it lacks distinct detail. The pleasures of this sort of history are formal and abstract, a bit like the aesthetic appeal that Adam Smith attributed to the contemplation of the workings (and wonders) of the market.

This sort of history can be contrasted with what I call 'refuge' history. Refuge history is close-up and on a small scale. Its emphasis is on a particular place rather than space, the careful delineation of particularities and details. The pleasures of refuge history derive from a sense of belonging, of connectedness – to both persons and details – in which the observer is also a participant. Its assumption is that knowledge and insight come from sympathy and understanding, from a process of loving recuperation. Refuge history is *heimlich*. It looks to the Adam Smith of *The Theory of Moral Sentiments* rather than *The Wealth of Nations*. The assumption is that, as ordinary people who lead ordinary lives, ordinary life in the past is something that we can relate to our own experience, and something that we can imaginatively inhabit. The distance between the past and our own experience is radically reduced, because we believe that we have the same sort of experience – that of the everyday. And the reduction of scale makes everything more human – easier for the viewer to relate to sympathetically.

A third sort of reenactment is what Robert Darnton has called 'incident history', and whose best example is probably the various versions – textual, filmic and musical – of *The Return of Martin Guerre*. Some of these – like *Martin Guerre* – belong to the old and time-tested genre of 'strange but true' – their appeal is because of their oddity (and by virtue of that fact belong in the ahistorical category of the not normal), but many others, such as the rehearsal of political crises or major battles, attempt to bring together the two sorts of history I have called prospect and refuge.[13]

But the general point I want to make is that almost all of these versions of how reenactment can and does work are shaped by the dominant terms sympathy and identity, whose chief virtue lies in the claim that they reduce distance. Now, while I see the merits of the claim for distance reduction, I see no merit to the notion of distance elimination. Indeed I see claims of distance elimination as inevitably and invariably false, illusory even when claimed by the re-enactor to be subjectively authentic. Distance, I want to suggest is something we should work with rather than seek to remove. Once we recognize that it inevitably is there, we can do other things with it apart from seeking its elimination.

All of this is by way of preliminary remarks before embarking on a case of show and tell. What I want to do is to look at how Roberto Rossellini recovers the experience of Italy's liberation in ways which do not occlude distance and which, in my view, render a far more plausible and realistic reenactment of what the process of liberation meant for many of those who participated in it. *Paisa* – together with *Roma, Citta Aperta* – is conventionally cited as one of the first and most important films called neo-realist. Neo-realism was not just a cinematographic convention, but part of a larger post-war cultural movement intent on ending the ideological and aesthetic obfuscation of everyday life, a process largely (though not exclusively) equated with the spectacular culture of Italian fascism. (Later, in the Cold War, its target shifted to the spectacle of Hollywood and of consumer society.) Though, as many of those identified as neo-realist have pointed out, not a coherent movement, it exemplified a quite shameless attempt to render the real. As Italo Calvino put it in the preface that accompanied the re-publication of his great neo-realist novel, *Il sentiero dei nidi di ragno* (first published in 1947): 'those who now think of "Neo-realism" as a contamination or coercion of the literature by non-literary forces, are really shifting the terms of the question: in reality the non-literary elements were simply there, so solid and indisputable that they seemed to us to be completely natural; for us the problem appeared to be entirely one of poetics, of how to transform that world which for us was *the* world into a work of literature'.[14] Rossellini said, 'What mattered to us was the investigation of reality, the [film's] correspondence to reality'.[15] The critic Cesare Zavattini, who was also the scriptwriter for *Ladri di biciclette*, put it similarly: 'What we are really attempting is not to invent a story that looks like reality, but to present reality as if it were a story'.[16]

Zavattini's neo-realist manifesto adduced three main principles for neo-realist cinema: to shoot in the present tense in order to explore the impact of history on everyday life; to use episodic narrative to historicize the role of chance; to use non-professional actors to show that the flow of public history and the rhythm of private behaviour are neither neatly joined nor totally unrelated, but mostly at odds, while their effects overlap and intersect with one another.[17]

Roberto Rossellini's *Paisa,* perhaps more than any other film, comes to embody these aims and principles.[18] I'm aware that there is considerable debate about the interpretation of the film, particularly about its contribution to issues of nationhood and identity at the end of war.[19] But here I'm chiefly concerned with Rossellini's strategies, both rhetorical and particularly narratological, in conveying or representing 'the real'. It is instructive, in this context, to contrast *Paisa* with Rossellini's other post-war masterpiece, *Roma Citta Aperta.* The latter has rightly been characterized as a melodrama, one in which there is a clear struggle between wrong and right. But *Paisa* takes an altogether different form, one that plays out the tension between a progressive heroic narrative of the liberation of Italy by the British and Americans in 1943–4 and a series of stories that cast an unflinching eye on the human costs of freedom. These stories are not figured as a struggle between right and wrong (there is brutality on all sides), but as a study in how the grand sweep of history gets worked out in everyday lives. Its themes are not bravery and heroism – though this is shown – but of shared sacrifice and suffering (a Christian theme found in much of Rossellini's work).[20] The characters are pushed and buffeted by the grand forces of the military struggle but also by chance, misfortune and misunderstanding. We are made acutely conscious of how strongly attached people are to the intimate pleasures of ordinary life – love, friendship, family, (Christian) fellowship – how these are sought as a comfort and a refuge, and how powerfully they are disrupted by the sweep of history. The object of the film is not to re-act scenes from the invasion/liberation but to convey to the audience the emotional experience of those years.

Rossellini depicts six scenes set in different regions of Italy which follow the Allied advance, beginning in Sicily, then moving to Naples, Rome, Florence, Emilia-Romagna, finally ending in the Po Valley. This grand narrative is represented as an impersonal inexorable process, signed by brief clips of maps, diagrams and a large abstract arrow, marking the movement of the allies and of time. These shots are accompanied by reassuring newsreel footage of retreats, victory parades and cheering crowds of the sort familiar to everyone who watched newsreels in Europe and America at the time. They embody a cheerful optimism which is reinforced by our (retrospective) knowledge that the good guys will and did win. They constitute both in their narrative form and scale an instance of prospect history.

But the stories – in the present tense – have a very different effect. They reduce the conflict to a human scale – indeed, they make it human – yet in doing so they undercut or, at the very least, rewrite the positive story of liberation. In all but one case the effects and unintended consequences of the Allies' liberation – the sufferings and misunderstandings that follow in the wake of war – are revealed as the larger narrative thread is entangled with a series of ordinary, personal everyday stories.

Every segment of *Paisa* deserves detailed treatment, but the constraints of space force me to confine my remarks to the very first episode, set at the beginning of the Allied advance through Sicily. It opens with the grand historical narrative, but the heroic bombast of the voiced-over Allied landing quickly elides into a scene of confusion, a babble of voices, a tissue of misunderstanding, and a powerful sense of the entanglement of grand history and ordinary lives, when American soldiers first meet the local Sicilian population. As in the rest of the film the characters (many not actors but 'ordinary people') speak in a babble of languages and dialects – American English, English English, German, Sicilian, Neapolitan, Roman, Tuscan and Venetian dialects of Italian, as well as the pure version of the language. The opening linear narrative of the film is disrupted by a whole series of trajectories and points of view.

The American troops take a young Sicilian girl, Carmela, to guide them through a minefield. They reach a castle tower, leave the woman with one of their number, 'Joe from Jersey', and continue their advance. Much of the episode is taken up with the attempts of Carmela and Joe to communicate despite their ignorance of the other's language. Joe admits his fears, talks of his job as a milk-float driver back home, and of his sister and her child. Carmela, mistakenly thinking he is showing her his wife or girlfriend, betrays a certain jealousy, and Joe, to show the family resemblance with his sister, uses a lighter to illuminate the photograph in his wallet. Rossellini then brutally cuts to a group of German soldiers who cry out as they see Joe's light. One shot rings out. Joe falls dead at Carmela's feet. The swiftness of the transition is shocking. Carmela hides as the Germans take over the tower. Angered by the death of Joe – his gestures of friendship have rendered him a friend not an alien presence – she shoots one of the Germans with his rifle. When the Americans return to the tower, they find Joe's body and assume that Carmela, 'the dirty Eye-Tie', has treacherously killed him. But in the final shot we see Carmela's body sprawled on rocks at the sea's edge where she has been thrown by the Germans.

This episode contains many of the features repeated in other parts of the film. The military action, filmed under cover of darkness, is hard to understand both for the soldiers and for the film's viewers (just as in the final episode set in the Po Valley, a night landing of supplies in almost completely unintelligible). Apart from the scenes with Carmela and Joe, the episode is shot at middle distance so that the spectators are not directly engaged in the action but watching an unfolding story whose plot is almost impossible to detect. Here as elsewhere in the film the camera conveys in its framing and movement what Leo Braudy has called a 'sense of detached, almost cold observation' that undercuts a sentimental view of events.[21] Again, as throughout the film, the longing for intimacy, understanding and friendship clashes headlong with the brutal pressures of war. Time and again moments of connectedness and warmth – a shared meal, the washing of hands, a gift of a medicine, playing a mouth organ – are disrupted by conflict; dreams of union are thwarted.

At one level Rossellini is offering us a universal account of the vicissitudes of war, a sort of humanist analysis of what military conflict means for those who cross its path. At the same time, in its sensitivities to place – each of the episodes has regional character and feel, conveyed by the landscape and the people who inhabit it – it is culturally specific. (Neo-realist commentators, regardless of their ideological complexion, were highly conscious of how what they were doing grew out of a special historical conjuncture.) But Rossellini is not maudlin and rarely sentimental. His audience, like many of the characters in *Paisa*, is on a journey of discovery or realization. It is not asked to act a part in the theatre of sympathy. The camera, the vehicle of image making – restless, mobile and detached, works not as a bond of sympathy but as a tool of understanding. Throughout his career, Rossellini was preoccupied with the persuasive and instructive power of the visual media – a huge part of his output were the numerous and lengthy historical programs he made for television in his later years.[22] In *Paisa* (and other work of this period) Rossellini seems concerned with making clear that we should learn from history, even when we recognize that we have moved beyond it. We can reconstruct our sense of the past without having to inhabit it, because the point of reconstructing the past is to make us better equipped to live in the present. As many critics have pointed out, this sort of position is very much derived from Benedetto Croce, the influential liberal Italian philosopher.[23]

In the *Poetics* Aristotle famously distinguished historical truth that was concerned with particular events and what did or has happened, from poetic truth that was concerned with universal (human) matters and with what might or what we imagine could have happened.[24] Some commentators on *Paisa* have been sought to emphasize its historical truth – the use of 'real' American soldiers and Italian partisans, or Neapolitan street urchins, and the filming in the 'real' locations where battles raged, and atrocities were committed. But every bit as important as these fragments of authenticity and veracity are to the film, they pall in my view when placed beside the much larger verities that Rossellini explores about conflicting temporalities, about the complex dialectic between big history and small everyday lives, and about the way in which the forces of history are constantly led to deviate or change as a result of chance. (This, crucially, is neo-realism's notion of the real.) Reenactment, it seems to me, is not going to get much beyond a site of modern fantasy and nostalgia (pleasant as this may be), unless it can begin to address the issues of the relationship between historical and poetic truth (I know Aristotle wanted to keep them apart, but never mind), the issue of forms of narration, and of dealing with contingency and chance. And to succeed in doing so one needs, as Rossellini's camera shows, a degree of detachment, a consciousness of the processes by which we represent the real.

Notes

1. A. Geppert (2004) 'London vs. Paris: Imperial Exhibitions, Transitory Spaces, and Metropolitan Networks, 1880–1930' (Ph.D. dissertation, European University Institute, Florence).
2. P. Nora (1984) *Les lieux de mémoire, sous la direction de Pierre Nora*, 7 vols (Paris: Gallimard).
3. H. Garfinkel (1967) *Studies in Ethnomethodology* (Englewood Cliffs: Prentice Hall). *The Ship* appeared on BBC in 2002. An account of the production appears in S. Baker (2002) *The Ship: Retracing Captain Cook's Endeavour Voyage* (London: BBC).
4. J. Lamb (2008) 'Historical Re-enactment, Extremity and Passion', *The Eighteenth Century: Theory and Interpretation*, 49:3, 51–62.
5. For an excellent account of the debate, see L. Re (1990) *Calvino and the Age of Neo-Realism: Fables of Estrangement* (Stanford University Press: Stanford), especially Chapters 1–3.
6. Available at http://genealogy.about.com/od/wars/a/reenacting.htm?terms=re-enactment.
7. Thus debates at websites concerned with 'standards of authenticity' seem chiefly preoccupied with dress and utensils, whether domestic or military.

8. Even the most perfunctory search of re-enactment websites reveals an overwhelming military predominance.
9. M. Jay (2005) *Songs of Experience: Modern American and European Variations on a Universal Theme* (Berkeley and Los Angeles: University of California Press), pp. 5–6.
10. K. Kosik (1976) *Dialectics of the Concrete: A Study on the Problems of Man and the World* (Dordrecht, Holland; Boston: D. Reidel), p. 2.
11. T. Downing (2004) 'Bringing the Past to the Small Screen' in D. Cannadine (ed.) *History and the Media* (Basingstoke; New York: Palgrave Macmillan), p. 17.
12. J. Appleton (1975) *The Experience of Landscape* (London: Wiley), p. 73.
13. R. Darnton (24 June 2004) 'It Happened One Night', *New York Review of Books*, 51:11. For some important comments on the appeal of such strange but true stories to historians, see C. Gallagher and S. Greenblatt (2001) *Practicing New Historicism* (Chicago; London: University of Chicago Press), pp. 54–6.
14. I. Calvino (1998) *The Path to the Spiders' Nests*, trans. A. Colquhoun, rev. M. McLaughlin (London: Jonathan Cape), pp. 9–10.
15. R. Rossellini, (1992 [1952]) 'A Discussion of Neorealism' reprinted in R. Rossellini, *My Method: Writings and Interviews*, ed. A. Apra, trans. A. Cancogni (New York: Marsilio), p. 41; D. Forgacs, S. Lutton, and G. Nowell-Smith (2000) *Roberto Rossellini. Magician of the Real* (London: British Film Institute), pp. 39, 151.
16. R. Stam (2000) *Film Theory: An Introduction* (Oxford: Blackwell), p. 73.
17. For Zavattini's views, see Cesare Zavattini, Valentina Fortichiari, Mino Argentieri, Gian Piero Brunetta (eds) (2002) *Cinema: Diario cinematografico Neorealismo ecc.* (Milan: Classici Bompiani), especially pp. 741–69.
18. Even though, it should be added, their views of neo-realism came to differ.
19. A. Dalle Vacche (1992) *The Body in the Mirror: Shapes of History in Italian Cinema* (Princeton: Princeton University Press), pp. 102, 180–8, 194–5, 197–201; P. Brunette (1983) 'Unity and Difference in *Paisan*', *Studies in the Literary Imagination*, 16, pp. 91–111.
20. For Christian neo-realism, see T. Gallagher (1988) 'NR=MC^2: Rossellini, "Neo-Realism", and Croce', *Film History*, 2, pp. 87–97.
21. L. Braudy (1978) 'Rossellini: From *Open City* to *General della Rovere*' in L. Braudy and M. Dickstein (eds) *Great Film Directors: A Critical Anthology* (New York: Oxford University Press), p. 659.
22. The best-known remnant of this work is his *The Rise to Power of Louis XIV* (1966).
23. See especially Gallagher, 'NR=MC^2', pp. 90–1.
24. Aristotle (1995) *Poetics* IX, trans. Stephen Halliwell (Cambridge, MA: Harvard University Press), pp. 59–61.

7
Textual Realism and Reenactment

Jonathan Walker

This chapter introduces an aesthetic strategy that bears a superficial resemblance to pastiche, but involves more than just mimicking the idiom and style of another writer or painter. I call this strategy 'textual realism'. It is a self-conscious and critically engaged form of intertextuality, whereby a modern film, novel or work of history incorporates references to and/or quotations from visual and literary sources dating from the period in which it is set. In other words, this strategy redefines the notion of realism and changes its target. Rather than attempt to represent directly the lived reality of a past era, it re-enacts the characteristic ways in which a past era represented its experience to itself.[1] The idea of textual realism thus provides a way to sidestep the fact that we cannot have direct, unmediated access to the past. It turns this alleged problem into an intellectual and aesthetic challenge.

What is the difference between textual realism and pastiche? According to Fredric Jameson's famous definition, pastiche is 'blank parody, a statue with blind eyeballs'.[2] It 'does not reinvent a picture of the past in its lived totality; rather, by reinventing the feel and shape of characteristic art objects of an older period ... it seeks to reawaken a [nostalgic] sense of the past associated with those objects.[3] It is 'an alarming and pathological symptom of a society that has become incapable of dealing with time and history'.[4] It substitutes 'art about art, images of other images' for an effective or critical engagement with history, so that 'the past as "referent" finds itself gradually bracketed, and then effaced altogether, leaving us with nothing but texts'.[5]

By contrast with the blank stare of pastiche, textual realism replaces imitation with selective, critical quotation and commentary – that is, with interpretation – but this interpretation proceeds through dramatisation rather than exposition: through reenactment.

Consider Tony Richardson's underrated film, *The Charge of the Light Brigade*, made in 1968 and set during the Crimean War in the mid-nineteenth-century.[6] Richardson is immediately faced with the challenge of authenticity. Is telling a story set in the nineteenth century by means of modern media (that is, moving pictures intended for projection upon a cinema screen) an anachronism? Most complaints on the issue of anachronism concern questions of content or mentality – the latter usually involving the attribution of modern attitudes and beliefs to historical characters. The idea of formal anachronism is rarely raised.[7]

For Jameson, the unwillingness or incapacity to acknowledge anachronism is one of the fundamental characteristics of pastiche. To put this in positive rather than negative terms, one of the achievements of pastiche is to actively suppress the concept of anachronism. By contrast, deliberate use of anachronism, and especially of formal anachronism, is the central feature of textual realism, in which the tension – even the contradiction – between modern methods of storytelling and the very different narrative techniques used by people in the past is a creative tension. The only unforgivable error would be to pretend that this tension did not exist–as pastiche does. History exists to map the fault lines between the past and the present, rather than to paper over the cracks.

The Charge of the Light Brigade is punctuated with animated sequences – made by Richard Williams – that are its most brilliant coup. These are very obviously not realistic at all, at one level. On the contrary, they consist of moving allegorical tableaux that dramatise relations between the European nation states (the English lion and bulldog, the French cockerel, the Russian bear). However, their style is realistic in the sense that it invokes the satirical cartoons from the magazine *Punch* or the etchings that Phiz created for Dickens' novels – and also perhaps their eighteenth-century forebears, William Hogarth and James Gillray. Considered as pastiche, the animations are lovingly detailed, and their tone faithfully reproduces the imperialist rhetoric of the mid-Victorian era. But they are not just pastiche. Something has been added to the original sources: most obviously, the simple fact of animation, but with it has come a different attitude, a kind of detachment and self-conscious manipulation of hindsight that is (by definition) absent from the primary sources. Very quickly, the integrity of the representation is (deliberately) undermined, as unified tableaux disintegrate into collaged fragments in a way that anticipates techniques later used by Terry Gilliam in *Monty Python's Flying Circus*. Anachronism is deployed as a critical technique.

This is especially apparent in an animation celebrating the 'Fall of Sebastopol', which is prefaced by a live-action sequence in which a character reads a newspaper account of this pseudo-event, and responds spontaneously by breaking into a waltz. From this we segue into the animation, which draws to a climax in a burst of fireworks, which in turn segues into a cannon salute, at which point the film jumpcuts back to live action in the Crimea, where a shell explodes above the local correspondent for *The Times*, William Howard Russell, who is taking shelter in a foxhole. The following exchange with the asinine Captain Duberly then reveals the falsity of the reports in the English press.

RUSSELL: Fallen me arse.
DUBERLY: (withdrawing newspaper from his uniform) It does say in *The Times* newspaper that Sebastopol has fallen.
RUSSELL: I am fallen.
DUBERLY: And I'm fallen alongside you!
RUSSELL: I'm not responsible for every damn lie that newspaper prints. You mustn't believe all you read in *The Times*, Captain Duberly. Sebastopol has not fallen. I'm not aware how *The Times* came by such information.

The need to read sources critically is emphasised by Russell's dismissal of his own newspaper's veracity. But *The Charge of the Light Brigade* goes further than this simple reiteration of a basic principle of historical interpretation. It draws our attention to the limitations of earlier, historic technologies of representation by placing them within a wider palette of modern effects. This creative use of anachronism works not only by juxtaposing the deluded triumphalism of the animation with the grim realities of the Crimean campaign (although, of course, the latter appear only in the form of another historical reconstruction), but also by playing with realism effects. That is, it emphasises the distance between lived experience and representation in any historical era, so that this distance becomes the real (we might say the allegorical) subject of the drama. The film draws our attention to what the sources do not say, and, more importantly, it emphasises what the sources can never say by virtue of their intrinsic formal properties as specific media, irrespective of the particular insights or the blind spots of individual historical actors.

Apart from animation, the other obvious reference to an historically specific medium in *The Charge of the Light Brigade* is to photography. The Crimean War was not only the first to involve specially deputised war correspondents, but also the first in which the army was accompanied by an official photographer, one Roger Fenton (who subsequently had a significant place within the early history of British photography).[8] In the 1850s, photography could not capture movement. Fenton could not take pictures of battles in progress, only posed, static shots, which were composed in ways that were indebted to painting. Indeed, Fenton had trained as a painter. Thus, most of Fenton's shots in the Crimea were portraits of officers taken in camp, behind the lines, and his most famous images – of *The Valley of the Shadow of Death* (which do not, in fact, show the scene of the famous charge) – show no immediate evidence of recent combat other than discarded cannon balls.[9]

We can interpret the closing image of *The Charge of the Light Brigade* in the context of Fenton's curiously muted and tactful body of work. The film's anti-climactic denouement begins with a short scene in which the commanding officers argue over who is responsible for issuing the fatal order for the charge; that is, they argue over what their own historical legacies will be. Their voices quickly mingle into an indistinguishable cacophony. We then cut to a shot of a soldier struggling to raise a fallen, wounded horse. Then we cut to a long shot of the slaughter on the battlefield in the aftermath of the charge (a scene that is never shown in Fenton's photographs). On the soundtrack, this panoramic image is accompanied only by solitary gunshots. From the implicit logic of montage, we assume that these gunshots signal the disposal of injured animals and – who knows? – perhaps even men. From here, we cut to a final, tightly cropped long shot of a single, prostrate – and unmistakably headless – horse. This appears initially as a photographic image, but it then morphs into what looks like an etching; that is, it re-enacts the process by which photographs were converted for the purposes of mass distribution in the 1850s (the technology of photo-reproduction being at that date less advanced than the technology of photography itself). Fenton's photographs may originally have been intended for publication by such means in the *Illustrated London News*.

This closing image – which was heavily criticised at the time of the film's release in 1968 as being bathetic and even nihilistic – invokes the compositional logic of Fenton's still images and simultaneously dramatises the process of their intended duplication and mediation, but the stasis here is not that of a formal, staged portrait. Rather, it is the stasis of death. The sounds of terminal gunshots dissolve, not into silence, which would be the only truly 'realistic' accompaniment for a still photograph or an etching, but into the buzzing of a solitary fly, a sound that signifies the absence of life more eloquently than silence could ever do. It is this barely audible, restless, mono-maniacal sound that accompanies the closing credits as they roll.

What Tony Richardson (and Richard Williams) did was to take contemporary forms of representation (the photograph, the satirical cartoon) and add something to them (hindsight, the sound of a buzzing fly) in order to highlight the questions they raise for a modern audience and to emphasise the distance between reality and representation.

If we move further back in time, from the photograph to the painting, the effect of mimicking this latter medium within a film is even more striking, notably in Stanley Kubrick's *Barry Lyndon* (1975), an adaptation of a nineteenth-century historical novel by William Makepeace Thackeray. Throughout this notoriously leisurely paced film, the actors are placed within the frame in a highly self-conscious manner, like models – or perhaps automatons (an eighteenth-century fad). Many scenes are shot in long takes and contain barely any movement or dialogue. Often, the compositions quote eighteenth-century paintings.[10]

One of the most representative scenes occurs at approximately 91:00 to 92:00. In it, the protagonist (played by Ryan O'Neal, whose blank, uninflected face becomes an asset in this context) meets his future wife at a gambling table. The ostensible subject is highly dramatic: high-stakes gambling, with an undercurrent of sexual tension. But as the camera moves in to isolate the faces of the two protagonists from their backgrounds, cross-cutting between them but very deliberately *not* providing a master shot in which they both appear, the soundtrack is reduced to a disembodied and detached chorus, whose mutterings and gasps at the turns of fortune in the game could theoretically be non-diegetic, since they do not issue from any visible mouths or bodies. And although the established context makes it clear that both protagonists are deeply involved in the progress of the game, their minimal actions and reactions do not support this presumption.

What is the point of this elaborately artificial method of filming? The original paintings that Kubrick incessantly references appear 'natural' because we accept that people do not move in paintings. We do not say that a painting is unrealistic because everyone in it is standing still. We accept the frozen aspect of a painting as a necessary condition for its existence. Immobility is an integral (a constitutive) aspect of the world inside a painting's frame. By contrast, we know that the cinema is both capable of and, in a sense, designed with the specific purpose of showing movement, and also capable of integrating the action thus depicted with spoken dialogue and other diegetic sound. Hence, freezing characters in near-silent tableaux within a film creates a feeling of unnatural constraint and tension. The reason it is so striking is that it is gratuitous. Kubrick could make his actors move and interact 'naturally', but he willfully chooses to impose restrictions upon them that mimic those under which eighteenth-century painters worked. He uses these restrictions to comment allegorically on aspects of eighteenth-century culture, and in particular on the difficulties that his protagonists face in negotiating their class boundaries.[11]

I was inspired by these examples – and in particular by the animation sequences in *The Charge of the Light Brigade* – to try something similar in a non-fiction narrative, the biography of a Venetian spy named Gerolamo Vano, who was executed in 1622 for perjury. My book about Vano is called *Pistols! Treason! Murder!* One of its most distinctive features is a set of original illustrations that serve as a visual commentary on both the events of Vano's life and on my retelling of them. They are clearly inspired by seventeenth-century woodcuts and engravings, notably those in Cesare Ripa's *Iconologia*, a standard emblem book of the period, and the Venetian costume books of Cesare Vecellio.[12]

To produce these illustrations, I worked with Dan Hallett according to a division of labour common in the modern comic book industry. In the terms of that industry, then, I am the writer and editor, and Dan is the penciller and inker. I do most of the initial picture research, and on the basis of that research I produce a written script and/or a storyboard. The precise arrangement of the compositions is then decided by Dan, who actually creates the images.

Figure 7.1 concludes the first chapter of *Pistols! Treason! Murder!*[13] It depicts a Dance of Death, which implies a direct continuity between myself (I appear at the end of the dance) and the protagonists of the story, whose order in the dance corresponds to that of their deaths in the historical narrative. The location is the liminal zone between the columns of San Marco: the site of both public executions and ceremonial entrances to the city in Venice. Hence, this space corresponds to the epistemological status of all the illustrations that follow, which are similarly situated in an 'in between' place. It is quite clear that they are not 'factual' in the sense that a traditional historical narrative is; but nor are they 'fictional' in the sense in which that term is usually understood. They are clearly identified as ambiguous; which is to say, they are ambiguous in a creative sense that stimulates the reader's imagination without causing epistemological confusion. This image – and those that follow – serve as allegorical bullet points, or revision notes, that summarise the argument of the chapters they accompany, although some of them are in between in a more literal sense in that they are independent

Figure 7.1 Dance of death

interludes, by analogy with *intermezzi* – the short sketches presented between the acts of many seventeenth-century Italian plays.

Figure 7.2 is a sample from one of these *intermezzi*, a chapter entitled 'Odd One Out', which is a very simple, three-page comic strip.[14] In it, I am placed in conversation with Vano and the reader is invited to spot the titular fake among a series of genuine quotations from Vano's surveillance reports. The intention is to parody one of the more clichéd conventions of romantic biography – that I identify so closely with my subject that our thoughts and words become one. If the joke works, then it does so by unexpectedly juxtaposing visual cues from religious iconography (the artificial deferential postures, the distorted relative scale of the protagonists) with the decontextualising strategy of Pop Art, the relevance of which is underlined by a paraphrase of a quotation from Roy Lichtenstein.

But there is more to this *intermezzo* than a simple joke. It also emphasises that Vano's spy reports do not tell a linear story, but are instead modular and repetitive in their structure. That is why it is possible to chop them up and reassemble them in such an aggressive manner. As isolated fragments – both here and in their original context – the

Figure 7.2 Odd one out

quotations are simple and emphatic, but they are virtually impossible to organise into a coherent narrative. Finally the way in which the religious visual motifs clash with the profane content of the dialogue is also an echo of Vano's startling and unexpected cynicism. In an age of sectarian fanaticism, he seems utterly indifferent to religion, and that is one of the most distinctive features of his writing.

By contrast to the stand-alone images in 'Odd One Out', Figure 7.3 is part of a sequence accompanying a long narrative chapter that summarises – to put it crudely – Vano's rise to power as Venice's unofficial 'General of Spies'.[15] The text of *Pistols! Treason! Murder!* makes frequent use of jumpcutting, an approach that is possible because of a distinctive

Figure 7.3 Gerolamo Vano's rise to power

design using multiple typefaces. This jumpcut approach is repeated in the structure of the narrative comic strips, a medium that is by its very nature discontinuous. However, we are not normally conscious of this discontinuity because writers and artists go out of their way to smooth over the transitions between panels. Here, by contrast, the transitions are deliberately ragged, and many violate visual continuity cues.[16] In fact, the sequence as a whole does not actually make much sense unless you have already read the written text first. The comic strips therefore take the principle of jumpcutting even further than the text does, speeding up the story by cutting it to ribbons, a kind of violence that is only possible because events that are already familiar to the reader are being replayed through a different medium. As in *The Charge of the Light Brigade* and *Barry Lyndon*, the images draw attention to what is missing from the written accounts upon which they draw, just as the written accounts draw attention to the gaps in the original archival sources.

This discontinuous style of narration – while it may be alien to the modern comic book – is quite common in seventeenth-century broadsheets, which (like my book) assume the existence of a linguistic commentary to accompany and to frame the presentation of the pictures. What the comics represent then is a sort of synthesis of modern visual narratives and earlier forms. Once again, they are 'in between', and they are doubly displaced by being presented in the form of a quasi-allegorical commentary upon the action, rather than as a straightforward retelling of it – again recalling the animation sequences in *The Charge of the Light Brigade.*

If the first narrative comic-strip sequence represented Vano's rise to power, Figure 7.4 is the centre piece of a strip that summarises his fall from grace, which was connected to the arrest of a noble named Antonio Foscarini.[17] The charges against Foscarini were not proclaimed publicly, which provoked a great deal of ill-informed gossip, a state of affairs that is dramatised in the illustration. The background is the Great Council Hall in the ducal palace, where the entire noble class met for debates and elections.

An argument that is never explored directly in the text of the book is dramatised visually in Figure 7.4. At the same time that the Venetian state

Figure 7.4 The arrest of Antonio Foscarini

was beginning to mount systematic surveillance operations targeted at individuals, Galileo was busy up the road in Padua, observing the surface of the moon and the satellites of Jupiter, and drawing some startling conclusions. By 1622, the year of Vano's execution, the first microscopes were circulating among curious cardinals in Rome. Moreover, the first question raised by Galileo's critics was the same one asked by Vano's readers: How can you be sure of what you have seen?[18] So there is an obvious connection to be made between the spy and the scientist. As Foucault would argue, power – in the form of surveillance – and knowledge – in the form of scientific observation – were intimately connected.[19] This argument is alluded to directly by the signature on the telescope at upper left, but it was already hinted at more obliquely in Figure 7.3, in which there is a visual quotation from Descartes.

The motif of the flies serves more than one function in Figure 7.4. Flies are not just examples of a preferred subject for early microscopic observations. They also refer to a linguistic metaphor introduced in a much earlier chapter. In the relevant passage I am addressing the reader directly in the portentous voice of The Historian.

> The living body does not exist for us, cannot speak to us, even if the corpse still hosts a different kind of life that has nothing to do with the consciousness that once inhabited it. Rather, this life is parasitical – a swarming mass of signs, continually multiplying, crawling across the page. Their buzzing is loudest around the body's wounds, where the text is most 'corrupt', as the philologists put it. The ligaments and cartilage that once articulated it have rotted away.[20]

This passage foreshadows a later throwaway comment about Foscarini's trial, in which 'No one ever originated rumours; no one confirmed or denied them. They were generated spontaneously, like flies in rotten meat'.[21] Figure 7.4 echoes all these previous allusions to flies. Finally, I suspect that these overdetermined insects are also direct descendents of Mosca, the buzzing parasite from Ben Jonson's play *Volpone*.

No doubt I'm already testing your credulity, but there is yet another argument implied by the contents of the other two telescope bubbles, in which the 'thing' being observed is actually a written text. This apparent paradox raises a point about the relationship between eyewitness testimony and hearsay and their respective evidential value in law – an issue that was crucial in the trial, condemnation and execution of Antonio Foscarini. The same point is also hinted at by the frieze of alternating eyes and ears, which have temporarily migrated to the panel border from Vano's cloak, where they normally reside (because Vano is not in control of the flow of information in this panel). Theoretically, evidence based on sight (the most noble of the senses) was of greater value that evidence based on hearing, which was frequently dismissed as mere gossip. However, in practice that distinction was virtually impossible to maintain, as the outcome of Foscarini's case demonstrates eloquently.[22] Again, this issue is not discussed explicitly in the text.

I want to turn now to a separate set of illustrations, which might be seen as a themed group of *intermezzi*, and which are represented here by Figure 7.5.[23] These illustrations accompany excerpts from a dialogue between me and two other historians. This dialogue is divided into

Figure 7.5 Dialogue

five separate sections, which intermittently interrupt the progress of the main narrative. In the distinct group of illustrations accompanying these dialogue chapters, the other two speakers are drawn as obvious caricatures, which represent particular historical schools: respectively political and diplomatic history, and social and economic history or – to put it in institutional terms – Oxbridge and a 'new university' in the 1960s or 1970s.[24] Their contributions to the written dialogue conform roughly to these characterisations.[25]

The dialogue chapters serve a number of functions in *Pistols! Treason! Murder!* First, they make the necessary exposition of background information both more informal and more dramatic. In effect, the dialogue excerpts form a secondary, episodic narrative that runs parallel to the historical events recounted in the book. Its five segments are presented as successive stages on a pub crawl, which the reader can follow on a map of Venice, and which in turn are implicitly juxtaposed with itineraries followed by seventeenth-century spies (thereby implicitly re-enacting them). This conceit dramatises a theme explored in theoretical terms in other chapters of the book; that is, the possible connection between walking and reading as interpretive strategies. Second, by setting the dialogue in particular spaces – in this case, bars – it was possible to challenge one of the most problematic conventions of academic writing: that knowledge presented in a neutral, disembodied voice speaks from nowhere in particular, and is therefore objective. The setting is in fact quite realistic, since many ideas on Venetian history really are thrashed out between colleagues in bars. Third, the dialogue is presented as though it were the transcript of a surveillance tape, a fiction that is signalled by a typographic cue: the use of a typewriter font, which continues to connote official documentation in the early twenty-first century, even though typewriters are now obsolete.[26] This apparently frivolous device of a tape transcript plays off the modern, one-way or hierarchical notion of surveillance against the Renaissance notion of mutual surveillance and display, epitomised by the older literary tradition of the dialogue, which assumes a free, open exchange between equals.[27]

The concept of the unreliable narrator is (or should be) intriguing for historians, as is clear from the animation sequence on 'The Fall of Sebastopol' in *The Charge of the Light Brigade*. The critical difference between historiography and the historical novel or film is usually assumed to be a matter of fact versus invention, but the question of sincerity is just as important in practice. Historians feel obliged to say what they mean. This restriction is more crippling to the imagination than that of factuality, and is not, I would argue, always essential. In the case of the dialogues in *Pistols! Treason! Murder!*, their content is ostentatiously implausible when considered as a record of an actual conversation. Again, this mimics the situation in Vano's surveillance reports, which similarly contain 'transcripts' of supposedly verbatim conversations that clearly never took place in the form they are presented. So there is a paradoxical situation in the dialogues, wherein the information presented refers accurately and honestly to real sources (unlike Vano, whose reliability is always in doubt), while the form in which that information is presented is (almost) utterly fraudulent. This fraudulence is clear despite, or even because of, the way in which the dialogue quotes clichéd typographic conventions to imply authenticity.

The style of the images for the dialogue sections also differs subtly from that of those that accompany the narrative chapters. The latter use a clear-line woodcut style to represent seventeenth-century subjects. For the pub scenes, by contrast, Dan included mock half-tone effects reminiscent of those used in 1960s and 1970s comic books (the period in which both the views of the protagonists and typewriter technology were current). However, the interpolation of protagonists from the opening Dance of Death, one of whom appears in each of the successive stages of the pub crawl holding a (clearly 1970s) tape recorder, makes it clear that we are still in an 'in-between' zone that is neither entirely historical nor entirely invented. Similarly each of the bar images includes figures from different, mutually incompatible historical eras in the background. In the case of Figure 7.5, outside the illuminated corner where the three protagonists of the dialogue sit, we find Jacob Burckhardt and Michel Foucault arguing vigorously over the meaning of modernity, while Casanova looks on from the background, no doubt admiring the barmaid's pose, which is borrowed from a Titian painting. The dialogue sections and their accompanying illustrations thus expand the idea of an 'in between' place, in which different pasts can meet on equal terms with different presents, without one historical moment having to displace any of the others. But unlike Jameson's description of pastiche, in which indiscriminate historical citation is a generalised condition, this place is a privileged one set off from its surroundings: a Carnivalesque space.

The comic strips in *Pistols! Treason! Murder!* and their stylistic antecedents in films like *The Charge of the Light Brigade* and *Barry Lyndon* raise the possibility of a new kind of reenactment, in which form is just as important as (if not more than) content. This kind of reenactment – which I have called textual realism – does not necessarily exclude either emotions or individual actors, but nor does it require them. Moreover it is not confined to individual experience (sensory or otherwise); nor is it confined to the replication of particular historical events or specific moments in the past. Instead, it creates the possibility of dialogue and exchange – literally, in the case of *Pistols! Treason! Murder!* Textual realism is reenactment as metaphor, so that the distance between then and now is simultaneously overcome and underscored. It is reenactment as commentary, in which the past can never be shown directly, but is instead represented by an allegory.

The experiences of Pierre Menard, a character in a short story by J. L. Borges, may help to explain the concept of textual realism. Menard began by attempting to re-enact Cervantes' composition of *Don Quixote* word for word, not by simply transcribing the text, but by remaking himself in such a way that Cervantes' words become the natural and inevitable expression of his own personality.

> Initially, Menard's method was to be relatively simple: Learn Spanish, return to Catholicism, fight against the Moor or Turk, forget the history of Europe from 1602 to 1918 – be Miguel de Cervantes. Pierre Menard weighed that course (I know he pretty thoroughly mastered seventeenth-century Castilian) but he discarded it as too easy. Too impossible, rather!, the reader will say. Quite so, but the undertaking was impossible from the outset, and of all the impossible ways of bringing it about, this was the least interesting.[28]

Menard's task therefore begins as a parody of an historicist ideal of reenactment.[29] However, he ends up somewhere else entirely. 'Being, somehow, Cervantes, and arriving thereby at the Quixote – that looked to Menard less challenging (and therefore less interesting) than continuing to be Pierre Menard and coming to the Quixote *through the experiences of Pierre Menard*'.[30] By a heroic (literally Quixotic) effort, Menard succeeds in composing a chapter and a few fragments of *Don Quixote* in exactly the same form that Cervantes did, but, as Borges' narrator points out, his achievement is quite distinct. The power of Menard's broken fragment derives precisely from its being taken out of context, both the context of the period in which Cervantes wrote and the context of *Don Quixote* itself. Similarly all history involves taking the past 'out of context'. This is unavoidable by virtue of the fact that we are talking about it in the present, quoting other peoples' words. All history is anachronistic. This is the paradox that both historicism and popular notions of reenactment fail to confront.

The story of Pierre Menard poses in a peculiarly exaggerated way the problem of the historian's relationship to sources. Historians do not create *ex nihil.* Nor does anyone else for that matter, but the constraints under which historians labour are severe and explicit in contrast to novelists and filmmakers. In *Pistols! Treason! Murder!*, I wanted to fully embrace these limitations – indeed, to highlight and to underscore them – and not to throw them off. Historians do not create stories from formless clay, but from material that has already been shaped by people in the past. We are magpies. This notion is a commonplace in the city I study, where the buildings are encrusted with loot from Constantinople. The Venetians stole the imperial heritage of Byzantium. They did not merely copy it: they literally broke it down, carried it away and put it back together again, making something new with it. I tried to do the same with the history of Venice, repaying the compliment they offered to the Byzantines.

In order to reuse and recycle our material, to say something new with it (and if we are not saying something new, then why say anything at all?), we must first take it to bits, gut it and expose the bare brick underneath. We must, in other words, violate its integrity by exposing its relativity, along with our own. But in order to perform this demolition safely, without danger to ourselves or to our readers, we must first thoroughly understand the internal structure of our sources. Only then can we reassemble them, incorporating the fragments into a narrative of our own. In other words, as T. S. Eliot once said of time, historicism is only conquered through historicism. A better programme than Menard's is suggested by another modernist, Picasso: 'Bad artists copy; good artists steal'.[31] I tried to steal the words of my sources, to appropriate or reinvent them while simultaneously fully acknowledging the original authors. In a twist on Menard's relationship with Cervantes, I wanted Vano's words to remain Vano's but simultaneously to become part of my book.[32]

Allegory, as Jameson puts it, 'consists in the withdrawal of its self-sufficiency of meaning from a given representation', which is to say, allegory explicitly invites contemplation of what a text cannot represent directly or adequately: in this case, perhaps, the historical referent, or 'the past in its lived totality' (Jameson again).[33] In the examples discussed above, the strategy of textual realism may resemble Jameson's definition of pastiche in some respects, but unlike pastiche, it does not depend on the suppression of anachronism. Rather it promotes anachronism as the central device for generating meaning.

Like allegory, anachronism puts the gap between past and present to use. In attempting to re-enact the past we inevitably get it wrong, but when we *deliberately* get it wrong, original insights emerge.

Notes

Thanks to Dan Hallett for permission to reproduce the illustrations.

1. Compare J. Forbes, (1997) *Les Enfants du Paradis* (London: The British Film Institute), pp. 20–1, on the philosophy of the set designer Alexandre Trauner, who worked on the titular film directed by Marcel Carné, released in 1945 but set in mid-nineteenth-century Paris. In creating the elaborate sets for *Les Enfants du Paradis*, Trauner rejected what he called 'primary realism' in favour of designs based on 'contemporary [i.e. nineteenth-century] illustrations, which are "already selections of reality"' or a 'second-level recreation', as Forbes puts it.
2. F. Jameson (2000) 'Postmodernism, or The Cultural Logic of Late Capitalism', in M. Hardt and K. Weeks (Eds) *The Jameson Reader*, Oxford: Blackwell, pp. 188–232, esp. p. 202 (originally published in 1983). For criticism of Jameson's definition of pastiche, see M. A. Rose, (1991) 'Post-Modern Pastiche', *British Journal of Aesthetics*, 31, 26–38.
3. F. Jameson, (1983)'Postmodernism and Consumer Society', in H. Foster (Ed) *Postmodern Culture* (London: Pluto Press), pp. 111–25, esp. p. 116. This passage does not refer to works with an ostensibly historical content, but rather to those that mimic obsolete historical forms – a point to which we shall return below.
4. Ibid., p. 117.
5. Jameson, 'Postmodernism and Consumer Society', p. 123; Jameson, 'The Cultural Logic of Late Capitalism', p. 203.
6. For a detailed description of the film, see M. Connelly (2003) *The Charge of the Light Brigade* (London and New York: I. B. Tauris).
7. I was (astonishingly) unable to find any extensive scholarly discussion of the concept of anachronism – as opposed to lists of particular errors contained in works of historical fiction, of which there are innumerable examples available online. This blind spot is all the more notable in the context of the discussion here, given that one of the most popular re-enactment groups in Europe and the US operates under the name of the 'Society for Creative Anachronism', and, moreover, the emphasis of this group's activities is not upon past events (battles, wars, etc.), but rather on customs and mentalities (i.e. upon the form rather than the content of historical experience). See www.sca.org.
8. R. Pare (1987) *Roger Fenton* (New York: Aperture Foundation).
9. Fenton's two photographs of The Valley of the Shadow of Death (and the precise nature of the relationship between them) are the subject of an inventive re-enactment by the filmmaker Errol Morris, who previously pioneered the use of dramatised re-enactments of events in documentaries like *The Thin Blue Line* (1988). In 'retaking' Fenton's static photographs, Morris emphasises the specific qualities of the technology available to Fenton and the circumstances in which he worked (chronological, topographical and historical). This is in contrast to previous interpretations, most of which are based on presumptions about Fenton's psychology and moral character. Morris' idea of re-enactment is therefore opposed to that of Collingwood and closer to my own as outlined in this chapter. See http://morris.blogs.nytimes.com/2007/09/25/which-came-first-the-chicken-or-the-egg-part-one/.

10. A number of specific visual comparisons are made in Pier Marco De Santi, *Cinema e pittura* (Date and place of publication unspecified: Giunti), pp. 58–61, which compares stills from the film to paintings by Gainsborough, Reynolds and Hogarth. The resemblances are striking and obvious.
11. One interesting consequence of this attempt to imitate an obsolete medium with a modern one was that it required a technological innovation. Kubrick wanted to shoot some scenes by candlelight, but the standard commercial lenses and film stock available in the early 1970s would not permit this. So he had to convert lenses designed by Zeiss for NASA, which were originally intended to photograph the dark side of the moon during manned space missions. One of the reasons the actors are so static is probably that the depth of field on a wide-open f0.7 lens is minute. Any movement by the actor would therefore require a focus pull, an extremely complicated procedure on the jerry-rigged cameras Kubrick was obliged to use to accommodate the lenses.
12. C. Ripa (1976) *Iconologia*, ed. S. Oregl, facsimile of the 1611 Padua edition (New York and London: Garland Publishing, 1976); C. Vecellio (1590) *De gli habiti antichi, et moderni di diuerse parti del mondo* (Venice: Damian Zenaro).
13. J. Walker (2009) *Pistols! Treason! Murder!: The Rise and Fall of a Master Spy* (Baltimore: Johns Hopkins University Press), p. 3.
14. Walker, *Pistols! Treason! Murder!*, p. 23.
15. Ibid., p. 61.
16. In the transition between the two full-page panels at ibid., pp. 168–9, the composition and point-of-view remain the same but the background and furniture change, while a new character also appears from nowhere. The result is a deliberate (and therefore productive) ambiguity as to whether a change of location has occurred between the two panels.
17. Walker, *Pistols! Treason! Murder!*, pp. 158–9.
18. M. Biagioli, *Galileo's Instruments of Credit: Telescopes, Images, Secrecy* (Chicago and London: University of Chicago Press, 2006). Galileo was both more capable and more willing than Vano to address this question directly and publicly. Nonetheless, despite Vano's limited abilities, and despite his failure to provoke a public debate or secure his own legacy (goals that preoccupied Galileo but that would never have occurred to Vano), there are interesting parallels between Vano's career and that of Galileo, as described in Biagoli's book.
19. M. Foucault (1991) *Discipline and Punish: The Birth of the Prison*, trans. A. Sheridan (Harmondsworth: Penguin).
20. Walker, *Pistols! Treason! Murder!*, p. 12.
21. Ibid., p. 148.
22. G. Buganza (1991) 'Il potere della parola. Le forza e la responsibilità della deposizione testimoniale nel processo penale veneziano (secoli XVI–XVII)', in *La parola all'accusato*, eds J. C. Maire Vigueur and A. Paravicini Bagliani (Palermo: Sellerio), pp. 124–38; J. Walker (2002) 'Legal and Political Discourse in Seventeenth-Century Venice', *Comparative Studies in Society and History*, 44(4), pp. 800–26.
23. Walker, *Pistols! Treason! Murder!*, p. 72.
24. My illustrator Dan refers to them as The Toff and The Geek respectively.
25. 'Phil' and 'Jim' have real life sources. The two characters are disgraceful misrepresentations of my friends Filippo de Vivo and James Shaw, with whom

I actually had a discussion on the topics raised in the dialogue sections of *Pistols! Treason! Murder!* However, I then rewrote the transcript of that discussion for dramatic purposes, and turned my interlocutors into fictional personae (with their generous consent).

26. See, for example, the 'Document Insert' sections in J. Ellroy (1995) *American Tabloid* (London: Arrow), which use the font American Typewriter. For the dialogue sections, *Pistols! Treason! Murder!* uses the digital font Typewriter P-22, which incorporates fake evidence of uneven key wear and impress into the letter forms – a further layer of visual artifice. I stole the idea of a dialogue with other historians from G. Corazzol (1997) *Cineografo di banditi su sfondo di monti, Feltre 1634–1642* (Milan: Edizioni Unicopli).
27. V. Cox, *The Renaissance Dialogue: Literary Dialogue in Its Social and Political Contexts, Castiglione to Galileo* (Cambridge: Cambridge University Press, 1992).
28. J. L. Borges (1998) 'Pierre Menard, Author of the Quixote', in *Collected Fictions*, trans. A. Hurley (London: Penguin), p. 91. The story 'Shakespeare's Memory' (in the same volume at pp. 508–15) is also a brilliant debunking of this mystical or Romantic version of historicism.
29. There is an unfortunate confusion about the term 'historicist' because Karl Popper used it to mean attempts to find pseudo-scientific historical laws – the sort of thing that doctrinaire Marxists used to do – while (even more confusingly) Jameson uses 'historicism' to mean 'the random cannibalization of all the styles of the past' in postmodern architecture (i.e. pastiche, according to Jameson's definition – see 'The Cultural Logic of Late Capitalism', p. 202). By contrast, I am using 'historicism' to mean interpreting the past on its own terms, using its own ideas and terms of analysis, and without imposing anachronistic judgements or terms of reference upon it. A classic statement of the principle can be found in Q. Skinner (1998) 'Meaning and Understanding in the History of Ideas', in J. Tully (ed.) *Meaning and Context: Quentin Skinner and His Critics* (Cambridge: Polity Press), p. 48: 'no agent can be said to have meant or done something which he could never be brought to accept as a correct description of what he had meant or done'.
30. Borges, 'Pierre Menard', p. 91 (emphasis in the original).
31. The phrase is apocryphal, and I have seen it attributed to others – notably to T. S. Eliot.
32. A more appropriate comparison might therefore be Eliot's use of direct quotation and pastiche in *The Waste Land* (1922), in which the citations and allusions depend on the reader's knowledge of their original context to obtain their full effect.
33. F. Jameson (1998) *Brecht and Method* (London: Verso), p. 122; Jameson, 'Postmodernism and Consumer Society', p. 116. The point below is raised to rather different ends in Jameson, *Brecht and Method*, p. 123.

8

'No Witnesses. No Leads. No Problems': The Reenactment of Crime and Rebellion

Paul A. Pickering

Opened in 1841, the foreboding buildings which comprised Melbourne's first penitentiary were extended several times until the completion of the bluestone walls and turrets in 1864. Although its design was inspired by the fashionable reformist ideas in penology of the day, the prison regime practiced within the heavy walls also incorporated a stark reminder of the lingering 'bloody code' that had characterised the British justice for centuries. Before it was closed in 1929 the prison had been the site of 136 executions. Among a grim list of offenders, the most notorious individual to die in the prison was Edward (Ned) Kelly. At 10:00 am on 11 November 1880, Kelly, his arms pinioned with a heavy leather strap, a white cloth bag folded back on his forehead, was led from a holding cell adjacent to the gallows. Outside a crowd estimated at between 4000 and 8000 had gathered. Although it had been over three years since an execution in Old Melbourne Gaol, it was not morbid curiosity which drew them there. These were Kelly's supporters (and opponents of capital punishment) who had hastily collected a petition bearing 30,000 signatures seeking a commutation of the sentence of death in the short interregnum between the court case and the day of execution. The numbers (neither crowd nor petition), however, could not disguise the fact that many Victorians supported the punishment of Kelly to the full extent of the law. Death by hanging was a popular verdict.

Upon reaching the 'drop', a noose was placed around Kelly's neck and the knot adjusted to ensure a quick death by the executioner, Elijah Upjohn, himself a prisoner recently transferred from Pentridge Gaol where he was undergoing sentence for stealing chickens. The prisoner had already received the last sacraments from the priest who had baptised him as a baby. The scene was austere, clinical, and officious. Before the cloth bag was pulled down to cover his face from the gawping onlookers below Kelly uttered his last words, his voice hushed, an admixture of defiance, resignation and trepidation. According to the reporters from the *Age* and the Melbourne *Herald*, Kelly's last words were 'such is life', although the rival daily, the increasingly conservative *Argus*, reported that he said: 'Ah well, I suppose it has come to this'. Others, closer to the action than the 27 official witnesses on that spring day in November 1880, claimed that Kelly's last words were entirely inaudible.[1]

Today Old Melbourne Gaol, as it is called, is one of the city's most successful tourist attractions. Much of the original prison has been demolished with only one of the original radial arms of its Pentonvillian design remaining. Everything you see is real, the visitor is told, and the seduction of the macabre objects is palpable: from an ominous frame for flogging to the hangman's toolkit; from Kelly's ghostly white death mask to the pistol he used in his infamous last stand with its handle chipped from the police bullet that shattered his little finger. Fortunately for tourist and tour operator alike the surviving part of the gaol contained the gallows which now provide the chilling centre-piece for an authentic chamber of horrors referred to disarmingly by the operators as a 'crime and justice experience'. The experience does not extend to standing on the 'drop' itself, the rope remains just out of reach. However, a clumsy miniature model in an adjacent glass case shows how it worked. Few resist pressing the red button.

Every Saturday, parts of the tempestuous life of Ned Kelly are re-enacted in the place of its ending (Figure 8.1). Written by Hannes Berger for the National Trust of Victoria, the pageant invites visitors to 'Step back in time and walk the road to the gallows in a 19th century prison'.[2] Visitors to the gaol, the playbill continues, will be given an opportunity to 'experience the truth behind Ned Kelly's life and legend'. Standing below 'the drop' Wendi Huggins, recounts the story from the point view of Ned's mother, Ellen, herself a prisoner in the gaol at the time of her son's death. 'Ellen' shakes her fist, points her finger accusingly, and in a wonderfully rich Irish brogue she excoriates the Victorian police, condemns the hanging judge, Redmond Barry, and bemoans the terrible fate that overtook her son. Pointing up at the gallows, tears welling in her eyes, she reports her son's final words. None of the ambiguity of the original scene is entertained. 'Such is Life' is not only the dénouement but also the title of the 'lively reenactment' played out at 12:30 and 2:00 pm every Saturday.

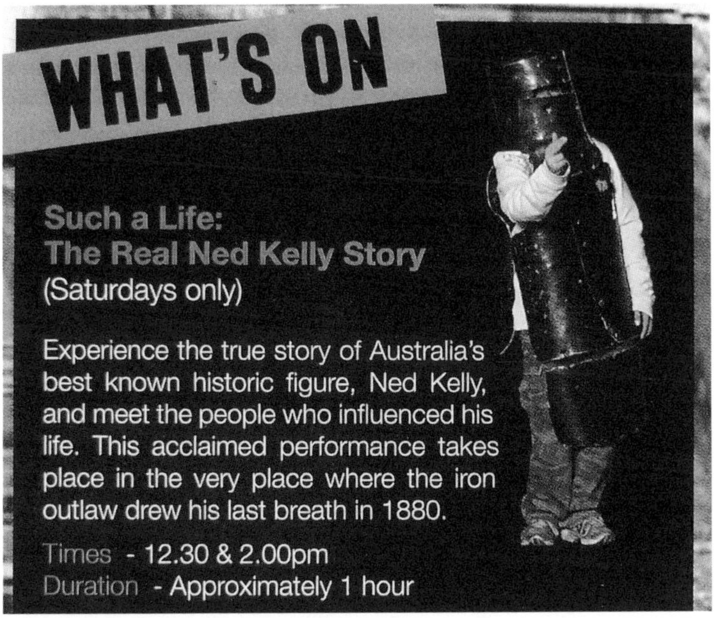

Figure 8.1 Old Melbourne Gaol publicity leaflet, Victoria, 2008

Two hundred and forty kilometres northward on the principal highway linking Melbourne and Sydney is the town of Glenrowan. Lying at the foot of the Warby Ranges, this small hamlet is home to less than 1000 people. The town's claim to a place in public consciousness is longstanding: it is best known as the location of a bloody gun battle which led to the capture of Kelly in 1880. To visit Glenrowan today the driver must leave the highway (this is true of all the towns that previously punctuated the route north). The turn-off is not easy to miss; within 20 kilometres large road signs, erected at the expense of the Victorian taxpayer, depict the outline of a helmeted-Ned brandishing pistols and proclaiming without any apparent sense of contradiction: 'WELCOME TO KELLY COUNTRY'. The Glenrowan townscape too is dominated by a menacing image of Ned in his infamous armour (Figure 8.2). In front of the main shopping centre a five-metre statue of an armoured Ned levels a rifle at all who approach, much as he did at the colonial 'traps' more than a century earlier.

Standing behind Ned's obdurate kerb-side vigil are several retail outlets including a souvenir shop, news agency-cum-post office, a Kelly museum and a tourist centre, Kellyland, the latter home to a multi-media

Figure 8.2 Main Street, Glenrowan, Victoria, 2008

reenactment unsurprisingly entitled *Ned Kelly's Last Stand*. 'Featuring 40 minutes of mind-blowing action', boasts the glossy publicity brochure available on the counter, 'THIS IS NOT A PICTURE THEATRE'. On the contrary, this show is a '5.5 million dollar extravaganza', a 'semi-live theatre production' featuring animation, commentary and 'computerised robots' which depict scenes in Kelly's miserable Hobbesian life in air-conditioned comfort 14 times a day.[3] The brainchild of local entrepreneur, Bob Hempel, who claims that his forebears forged Ned's armour in nearby Springhurst, the show makes an equally bold claim: 'The show is educational, historically correct and entertaining'.[4] 'Don't tell me that at some time in your life you haven't fantasized about going back into history and participating in an historical event', Hempel writes, '*Well here is your chance to live that dream*'. 'For through the brilliance of animation and computerised robots', he continues, 'you will be transferred back in time, over 100 years, to witness the events that lead up to the capture of the Kelly gang ... Starting as hostages in the hotel, and then onto gunfights – burning buildings – shoot outs – a decent hanging and finishing in our magnificent painting gallery'.[5]

Fifty kilometres north east of Glenrowan in the Victorian high country is the town of Beechworth. Founded in 1852 the town flourished during the Victorian gold rushes. Conspicuous consumption reached its apogee when, in November 1855, a local politician celebrated his election to the Legislative Council by riding a horse shod with golden shoes through the main streets of the town. The halcyon days were over long before Ned Kelly became a familiar figure on the streets in the 1870s. In 1870, as a 17-year old, Kelly served six months in Beechworth prison and the following year, having been at liberty for just three weeks, he was sentenced to three years' hard labour in Beechworth Courthouse for horse stealing. Beechworth was also the scene of his famous bare-knuckle fist fight with Isaiah 'Wild' Wright, the man Kelly blamed for his incarceration. Following his capture in Glenrowan, in August 1880, Kelly faced his preliminary hearing on the charge of murder in Beechworth's Courthouse.

According to the local tourist information centre, Beechworth 'is a blend of culture, adventure, mystery, history and passion – all the elements of a good novel'.[6] Preserved serendipitously in the clear air of the Victorian Alps, more than 30 local buildings have been classified by the National Trust, and much of the townscape has been lovingly restored to its nineteenth-century glory. It is an ideal site for reenactment. Every year a 'Golden Horseshoes' festival is held at Easter, featuring a parade, a reenactment of the bizarre peregrination of the ostentatious local representative and a hunt for a golden egg. The main attraction for reenactors in historic Beechworth is, however, undoubtedly the hapless Ned. For a number of years Beechworth has staged a reenactment of parts of Kelly's 1880 committal hearing and his Melbourne trial. In 2007 the courthouse drama was the centre of a Kelly weekend which included a reenactment of Kelly's transport from the railway station to the courthouse, an art show, site tours, shop window displays, Irish dancing and music, footage from early Kelly movies and public lectures by Kelly historians. Sandwich boards on the streetscape echo the main theme entreating passers-by to consume themed products: Ned Kelly Beer ('Such is Beer', runs the advertisement) and Ned Kelly Pies (made inexplicably from a mixture of steak, eggs, bacon and cheese). 'At the fully restored courthouse', promises another advertisement for tourists, 'you can stand in the dock and in the tiny cell where [Kelly] was held'.[7] Participants in the reenactment itself enjoy a special celebrity. Playing the part of Kelly in 2007 was local tour guide, Michael Beattie. Sporting a Kelly-like beard (grown to protect against the Beechworth cold), he looked the part. The whiskers were his, reported Melbourne's leading newspaper, the *Age*, but everything else belonged to Kelly. The 'stance', 'proud, righteous, determined', Beattie told the *Age*'s reporter was 'all Kelly's'. The reporter did not need much convincing, heading her report with a remarkable claim: 'Ned Kelly comes to life in Beechworth'.[8]

A common thread between these diverse examples of the reenactment of an iconic life in Australian history is the pursuit of profit. Long before Theodor Adorno lamented the ubiquitous commodification of culture, Ned Kelly had become a commodity. In 1906 Kelly's exploits were the subject of the world's first feature-length moving picture show. After a successful premiere in Melbourne – the press of the crowd necessitated a late change of venue to the Town Hall – and a run of seven weeks, the film went on tour through Australia and New Zealand in 1907 covering 11,000 miles and involving 150 screenings.[9] The box office receipts have also earned the film a place in history, if less well known. The film's producers recouped £25,000, a tidy profit on their outlay of £1000, a record percentage profit for any Australian production that still stands.[10] In an instance of terrible irony perhaps the most conspicuous beneficiary from Kelly's memory is the State that persecuted and executed him. Several studies have emphasised his benefit to the tourist industry of the State he opposed. It is estimated that the Kelly reenactment weekend in 2006, for example, brought 4200 visitors and over $1 million to the Beechworth economy, considerably more than he ever stole there. [11] Similarly historical reenactment theme parks would not survive without him. According to the Heritage Council of Victoria, Old Melbourne Gaol enjoys 140,000 visitors annually, generating revenue of over $1.15 million. Significantly it is the only one of the National Trust's 50 properties to make a profit, an outcome that is attributed directly to its close association with Kelly.[12] Walk into the souvenir shop at the entrance and it is perfectly clear who is paying the bills. The shelves are full of Kelly kitsch: from Kelly posters and key rings to Kelly dolls and even replica keys to his cell. Apparently no one owns the rights to Kelly's name and image, and consequently his memory is constantly being associated with products and services that bear no ostensible relationship to his biography and are nothing more than a flimsy pretext for the pursuit of profit.[13] The nadir of what is nowadays called 'Kellyana' is undoubtedly the Ned Kelly 'bunnykin' figurine produced by the Royal Doulton china company. The discordant image of a cute bunny rabbit armed with a carbine, ears protruding through the top of Kellyesque helmet, would surely be Adorno's nightmare.[14]

On top of this congeries of trash are attempts to re-enact Kelly's life through culture: music (from highbrow to muzak),[15] literature and visual art. Since the original film the Kelly saga has (to borrow the words of one critic) been 'inflicted on a long-suffering public innumerable times'.[16] In fact, since 1906 there have been ten films. The most recent, by Gregor Jordon, was released in 2003 and boasted major international film stars, Heath Ledger, Geoffrey Rush and Orlando Bloom.[17] Teetering on the brink of an oxymoron the producers promised to tell the 'true story' of a 'legend'. Following a run-in with the Commonwealth Censor, the film's distributor, United International Pictures, warned that an 'adults only' classification would 'prevent schoolchildren from studying the bushranger's story'.[18] In his introduction to a published edition of the screenplay, the film's producer, Jordan, professed his determination to get the story right,[19] but this did not stop him inflicting more violence on the facts than Ned visited on north-eastern Victoria, including giving the outlaw a most unlikely romantic interest.[20] When pressed Jordon admitted that he had acted to broaden the film's appeal with one eye firmly on the box office – the demographics – and the other on pleasing his investors.[21] The Commonwealth Censor, it seems, need not have worried about denying schoolchildren an educational opportunity.

As well as the films have been a bevy of plays. The first, *Catching the Kellys*, opened in Melbourne in March 1879, and a second, *The Capture of the Kellys*, opened in Sydney in July 1880, before the siege at Glenrowan.[22] The Colonial Secretary closed the latter play after a short run causing one disappointed prospective patron to offer £50 for the show to continue. 'Plendy munny!', he was reputed to have yelled, 'I want to see [the] Kelly gang!'.[23] He was likely to have been disappointed either way. By 1900 there had been at least a dozen stage productions including one which reputedly featured Ned's sister, Kate Kelly, riding his horse (the closest the patrons got to an actual encounter with the gang).[24] More recently, there has been a surfeit of television documentaries: at least half a dozen since the 1950s.

Jordan's effort to dress his film up as an educational experience, however, is a reminder that many Kelly reenactments lay claim to a higher purpose. The envelopment of Ned Kelly in a fog of commercialism, commodification and tourist dollars does not disqualify him as a subject for inquiry into the uses of reenactment or its efficacy as a tool for historical practice. On the contrary, his ubiquity in Australian culture – as hero or villain, as bank robber or bunnykin, as cop killer or cash cow, makes the multifarious reenactments of his life ideal vehicles for study. Among many possible avenues that suggest themselves, two threads will be explored in the remainder of this essay: the reenactment of Kelly on the screen – large and small – and the pedagogical and political issues raised therein, and the place of reenactment in the criminal justice system. Apart from Kelly himself, the link between these strands is the overarching aim of the chapter to contribute to both theorising and historicising reenactment and to better understanding its potential as a heuristic methodology.

A discussion of the representations of the Kelly saga on screen is given extra significance by the work of noted film historian, Robert Rosenstone. In an important book and a series of articles, Rosenstone argues that historical films must be 'held accountable to certain standards' which are 'consonant with the possibilities of the medium': 'It is impossible to judge history on film solely by the standards of written history, for each medium has its own kind of fictive elements'.[25] At the same time he has issued a fundamental challenge to historians, proselytising for what he calls the 'postmodern' potential of the historical film to recount the past self-reflexively and from a multiplicity of viewpoints. Where historians, even postmodernist ones, have failed, Rosenstone insists that filmmakers can 'create a past that is at once serious, complex, challenging, and "true" in its ability to render the meanings rather than the literal reality of past events'.[26] This claim could be extended to many other forms of reenactment. It is worth asking, therefore, do any of the Kelly reenactments – films and television docudramas – live up to Rosenstone's promise?

Films, such as Gregor Jordan's *Ned Kelly* which are driven, in the final instance, by box office imperatives, clearly do not. Based loosely on Robert Drewe's novel, *Our Sunshine,* the plot revolves around a titanic struggle between Kelly (Heath Ledger) and Superintendant Francis Hare (Geoffrey Rush). Writing in *Meanjin* in 1953, A. A. Phillips saw Douglas Stewart's play, *Ned Kelly,* in similar terms, an antipodean encounter between Ulysses ('the give-it-a-goer') and Telemachus ('the wowser') that heralded the arrival of Australian romanticism.[27] In the 1950s Phillips felt that Kelly's tale lacked the patina of age; nearly 60 years on, the same is true, or perhaps the protagonists are simply not strong enough to sustain an epic encounter. [28] Either way it does not work.[29]

The fact that Jordan was brazen enough to claim a role in education is a reminder of the myriad pedagogical uses to which reenactments (broadly defined) are put. This challenge has snuck up on the academy from behind. At one end of the spectrum is a branch of what is called experiential education or active learning.[30] Under this rubric school children routinely go on site visits where they play with objects and dress up in costumes in search of empathetic understanding with little reflection – by their teachers – on the nature of such knowledge.

Alongside sites of historical tourism (such as Old Melbourne Gaol, Beechworth Courthouse and Kellyland), are a plethora of historical films, television docudramas and documentaries. This is a corpus more than adequate to keep the *History Channel* on air around the clock and offers an educational experience that fits neatly into school curricula and caters to the increasing reliance on technology in the classroom. Much of this product is consumed without quality control: how many schools groups went on an 'educational' excursion to see *Ned Kelly*? The battle is all but lost. Published in 2003, a survey conducted by *History Today* showed that many university undergraduates in Britain have never read an entire book, taking all their knowledge of the past from the screen: computer, television and cinema.[31] Many university educators insouciantly dabble with 'affect' by uncritically using reenactments in their courses (who doesn't show the occasional DVD?) and some have even begun to explore online 'immersion' as a pedagogical tool, as can be seen by the enthusiastic use of the 'Second Life' virtual reality website by the Harvard Law School.[32] At least they are considering the implications.

Scripted under the watchful eye of noted Kelly historian, Ian Jones, the other modern blockbuster is Tony Richardson's *Ned Kelly*, released in 1970. This film is remembered primarily for the controversial casting of Mick Jagger to play Kelly, but the *enfant terrible* of the 1960s youth rebellion in Britain was actually an apposite choice (despite the inadequate whiskers and appalling Irish brogue).[33] The decision to film the story in New South Wales rather than in 'Kelly Country' in Victoria bruised local sensibilities and unnecessarily raised the question of authenticity. Apparently in a decision that would have resonated for Jean Baudrillard, Braidwood looked more like Glenrowan than Glenrowan.[34] For students of reenactment, however, Richardson's film raises the issue of politics, evident, in particular, in the existence of alternative endings. The (unexplained) alternative endings radically alter the film's message and its affective force. One cut of the film ends with Kelly's execution – a sombre, pathetic scene – which evokes predictable tropes: crime does not pay, and thus ends all rebellion. The second ends (approximately six minutes earlier) with Kelly's defiant courtroom exchange with Judge Barry (portrayed by the bulbous-headed Frank Thring). Thring, the face of Pontius Pilot for many filmgoers at that time, condemns Kelly with ritual pomposity; Jagger, the face of 1880s and 1960s youthful rebellion combined, points downwards and with a wry grin intones Kelly's famous retort: 'Yes, I shall see you there where I am going'.[35]

The reason why there are two diametrically opposed endings to Richardson's film remains a mystery, but it must be said that this is not the first time that a Kelly film has been changed.[36] On the contrary the paean of political dissent, evident in the shorter 'defiant' cut of the film, has been the cause of manifold legal troubles that have dogged Kelly films and plays alike.[37] As early as 1828 the New South Wales authorities legislated to control places of public entertainment ensuring that they were conducted in an orderly fashion and that patrons were not at risk of physical or moral injury. In 1850 these controls were augmented by the power to prohibit any entertainment to ensure the 'preservation of good manners, decorum or the public peace'.[38] From the outset, depictions of lawlessness – bushranging in Australian parlance – were at the same time popular with writers and audiences and deeply troubling to the government. Written in New South Wales, David Burns' *The Bush Rangers*, was performed in Edinburgh in 1829 but not on home soil for 142 years. Charles Harpur's *Bushrangers*, penned in 1835, was not performed in his lifetime. Writing in 1867 he was in no doubt why: 'a certain *soreness* on the subject of convict bushranging [...] existed in some rich and influential sections of colonial society'.[39]

In his pioneering study of social banditry, Eric Hobsbawm identified Kelly as an antipodean example who met the essential characteristics.[40] 'The point about social bandits', Hobsbawm tells us, 'is that they are peasant outlaws whom the lord and state regard as criminals, but who remain within peasant society, and are considered by their people as heroes, as champions, avengers, fighters for justice, perhaps even leaders of liberation'.[41] Despite Kelly's unpopularity with many Victorians, the venerable Judge Barry was clearly troubled by the breadth of his appeal: '[T]here is unfortunately a class which disregards the evil consequences of crime', he fumed when sentencing Kelly to the gallows, 'Foolish, inconsiderate, ill-conducted, unprincipled youths unfortunately abound, and unless they are made to consider the consequences of crime they are led to imitate notorious felons, whom they regard as self-made heroes'.[42]

Referring to the first Kelly film, the *Bulletin* was in no doubt that the nascent genre would be as popular as beer. Notwithstanding their strident advocacy of Australian nationalism, the editors were as concerned with the medium as they were equivocal about the message. Despite a grudging acknowledgement of film's potential as a means of mass communication and entertainment, they were certain that, like alcohol, it would be bad for its consumer:

> There is a deal too much racket in connection with this show – sometimes you can't see the pictures for the noise of horses, trains, gunshots, and wild cries; but, all the same, it is the sort of bellowdrama [sic] that the lower disorders crave for, and two-thirds of Australia will want to see it – the two thirds that believes Ned Kelly was a greater man than George Washington.[43]

Not surprisingly, most States and subsequently the new Commonwealth government included a specific clause in the legislation governing censorship to control films about bushrangers. For example, clause (iv) of the New South Wales Theatre and Public Halls Act (1908), later copied by the Commonwealth, prohibited the representation of 'successful crime, such as bushranging, or other acts of lawlessness'.[44] The impact on children was particularly mentioned in the Act. Earlier the *Bulletin* had reported that the screening of a Kelly film had provoked a juvenile crime wave in Ballarat. In a precocious Kelly reenactment, five 'kids not yet in their teens', broke into a photo studio and stole £8 10s and next day, armed with loaded revolvers, they 'bailed up some school kids in the best Ned Kelly style while other comic opera desperados went through the kids' pockets'.[45] As Ina Bertrand has noted there was a widespread concern about the effect of films on social behaviour, both inside the theatre (after all they were screened in the dark) and outside where the effect of realism would surely be imitation.[46] Viewers would, like the patrons of Kellyland in Glenrowan, be inspired 'to live that dream'.

The potential for mass communication meant that inevitably films about bushranging touched a raw nerve. Although the first tour of the film was relatively unaffected by official action (screening was only restricted in north eastern Victoria, 'the scene of some of the Kelly exploits'), a second tour in 1912 resulted in bans in South Australia, New South Wales and Victoria. The outcome varied from State to State but versions of the film released in the 1920s and 30s were either heavily censored or banned. Harry Southwell's *When the Kelly Gang were Out*, for example, was prohibited in 1923 despite the late inclusion of a still caption as a coda to the action. It read:

> And such is the inevitable fate of all who rebel against the righteous and mighty forces of law and order.
>
> Even so – better that scaffold than being hunted like dogs – what fools we must have been – what fools![47]

A second application to show the film in 1925 was also refused despite the removal of 'certain objectionable scenes'. A similar fate awaited a 1934 version of the story, again by Southwell, which was not released until 1943 because it 'glorified bushranging'.[48] The alternative endings to the Richardson film have more than a faint echo of these earlier troubles. Despite the fact that Ned's image in popular culture has been abstracted beyond all recognition, it would be a mistake to underestimate the lingering political undercurrent and the trouble it may cause.

Although they open up interesting areas for discussion neither Jordan's nor Richardson's feature films live up to the promise held out by Rosenstone. The gulf between a reenactment that inevitably contains fictive elements arising from decisions taken in the course of narrative construction and a story that self-confessedly takes liberties with poetic licence is immense. Nevertheless for students of reenactment there is much still to learn from an examination of representations of Kelly's life on the screen. A 2000 reenactment of Kelly's trial broadcast on the State-Funded Radio National, for example, spawned a formative interactive segment, 'Was He Guilty?', on the leading commercial network's flagship current affairs programme, *60 Minutes*, and a string of articles.[49] The line between informed public debate and popular entertainment was blurred. By inviting the audience to vote, Channel Nine were, whether they knew it or not, toying with what has since been called the 'democratic potential' of television.[50] The experiment with participation in the 'verdict' of the trial (re-trial) of an iconic popular figure was an innovation anticipating *Big Brother* and the generic music *Idol* series by years.[51]

Other 'trials' of Kelly are equally interesting for the historian to cogitate on. A television play by Roger Simpson, broadcast in 1977, was set in court, but not exclusively so. To embellish the proceedings Simpson employed a Narrator who physically intervenes in the action, berating the trial judge, and giving Kelly a voice that he was supposedly 'denied' in 1880. It also included a reenactment of the events at Stringybark Creek on 25 October 1878 when the Kelly gang 'bailed up' and murdered three policemen, interspersed at the appropriate point in the narrative. The reenactment was intended to aid Ned's defence. 'As a witness to the "actual" death of Lonigan', wrote one commentator, 'the viewer is enabled to sit with the Narrator in judgement on the trial itself'.[52] Most commentators agree, however, that Ned's claim that he acted in self-defence is better aided by broader contextual evidence – including the fact that the police had a licence to shoot him on sight. By its very nature this evidence is not included in the reenactment. Without context the reenactment of the shooting does Kelly few favours.[53] The same is true of the most comprehensive reenactment of the Kelly tale on film, the four-part mini-series produced in 1980 to coincide with the hundredth anniversary of Kelly's execution: *The Last Outlaw*. Although the vision looks passé only 30 years later, there is much here to make the professional re-enactor nod approvingly. 'Everything from location, buildings, and even the position of a swag on a saddle', boast the producers (including Ian Jones), 'has been conceived with a sense of getting it right'. Purporting to combine 'art with fanatical accuracy', the series uses its 397 minutes to present an extended justification of the Kelly uprising. Jones regards the 'Outlaw' as a putative revolutionary, 'perhaps' the 'only real Robin Hood who has ever lived' (the note of uncertainty relates to Kelly's uniqueness rather than the nature of his mission).[54] The sympathetic portrait, however, breaks down irrevocably at Stringybark Creek which looks like what it was, a brutal murder.[55] Despite their intended purpose these reenactments have a life of their own.

Unlike *The Last Outlaw* the most recent docudrama to essay the Kelly saga is deliberately polemical, stridently telling both sides of the story. Narrated by well-known film actor Jack Thompson, *Outlawed – The Real Ned Kelly* was screened on the national public broadcaster in 2003.[56] The 55-minute programme is framed by pejorative language – was Kelly a 'revolutionary hero' or a 'murderous thug' asks Thompson gravely – and the drama is cleverly heightened by interviews with descendants of both the Kellys and the slain policemen. This is not a film *à la* Rosenstone (nor does it purport to be), but it does at least attempt to complicate the dominant exculpatory narrative. The documentary is also notable for its use of Australia's 'number one' criminal profiler (and retired policeman), Stephen Longford, who ponders Kelly's so-called manifesto, dictated at Jerilderie in 1879. Although Longford's credibility as an expert witness in an actual court case at the time was questioned by a Supreme Court Judge,[57] junk science, as it is called, is clearly winning in the court of public opinion.[58] A recent on-screen attempt to 'find' Jack the Ripper, for example, was underpinned by the use of a criminal profiler and a retired detective from New Scotland Yard. This is surely an attempt to cash in on the popularity of the generic crime scene investigation programmes, predicated on the popular belief that historical 'truth' is a puzzle that can be solved with the aid of forensic science. As the profiler from New Scotland Yard, Laura Richards, puts it, 'a picture is worth a thousand words, a crime scene is worth a million'. Most of the programme is filled by Richards wandering through a cityscape long since changed out of all recognition, interspersed with reenactments of the heinous crimes committed 100 years earlier. The fact that they failed to uncover Jack's identity hardly seems to matter.[59]

Not surprisingly, many historians recoil reflexively from the idea of reenactment as either an irretrievably comical eccentricity or 'dangerous tosh'.[60] Greg Dening's oft-quoted dismissal of the entire movement as 'hindsighted superiority' leading to an hallucination of a past as, 'the present in funny dress', hangs in the air like Damocles' sword over the head of any historian willing to at least begin by taking it seriously.[61] A recent controversy highlights the critical issues more starkly than either Ned or Jack. In 2005 Kate Grenville published an historical novel, *The Secret River*, to popular and critical acclaim. Some commentators suggested that the book offered a new understanding of the history of early Sydney, and Grenville herself came close to suggesting that she had produced a new kind of history.[62] This aspect of the book's reception provoked a hostile reaction from historians. Inga Clendinnen, one of Australia's foremost ethnographic historians, described *The Secret River* as 'anti-history'. 'We cannot post ourselves back in time', she argued in a swingeing polemic, 'People really did think differently then'.[63]

In response to her critics, Grenville penned a detailed account of the 'research' that went into her novel. Historians will chuckle at the admission that she abandoned the Public Record Office after just one day in favour of a walk along the bank of the Thames (to 'look in the places where the past happened').[64] Returning to Australia empty-handed, Grenville went sailing up the Hawkesbury River, tried to make fire with sticks, and used an early settlers' slush lamp to illuminate her hut. The vast amounts of acrid smoke created by the latter proved seductively inspirational: 'In the next thirty seconds', she wrote, 'I learned more about life i n a bark hut on the Hawkesbury in 1817 than all the books in the world could have told me ... I went back to the desk and rewrote all the scenes in the hut at night'. Grenville has characterised this method as an 'experiential theory of writing'.[65] Although the word was not mentioned in the debate, what Grenville described was the use of reenactment. Her goal was affect, but the acrid smoke did nothing to narrow the distance between past and present. Nevertheless if we peek around the 'affective turn', we will find not only self-important novelists, but also lawyers and criminal investigators. Bypassing media-hungry retired detectives and profilers, historians can learn a lot by examining the use (and problems) of reenactment in the criminal justice system.

The intellectual crucible of reenactment as a methodology is the work of the Oxford philosopher, R. G. Collingwood. In his famous lecture series on 'the idea of history' presented in the 1930s, Collingwood devoted a whole section to what he called 'history as reenactment'. 'If the historian', he asked rhetorically, 'has no direct or empirical knowledge of his facts, and no transmitted or testimoniary knowledge of them, what kind of knowledge has he: in other words, what must the historian do in order that he may know them?' 'My historical review of the idea of history', Collingwood continued in a well-known formulation, 'has resulted in an answer to this question: namely, that the historian must re-enact the past in his own mind'.[66] Significantly Collingwood recommended as a model the fictional detective. 'The hero of a detective novel is thinking exactly like an historian', he writes, 'when, from indications of the most varied kinds, he constructs an imaginary picture of how a crime was committed, and by whom'.[67] Collingwood was an avid reader of crime fiction, but he might have just as easily looked to the burgeoning science of criminal investigation.[68]

Reenactment has a longstanding place in the law, both inside and outside the courts. By the 1930s, the potential of deductive reasoning and reconstruction seemed almost limitless: by this time the hero of the detective novel was thinking like a real detective. As the author of one textbook, Norman Kendal, put it in 1934, prior to Collingwood's lectures, the investigator must:

> reconstruct the occurrence, build up by hard labour a theory fitted in and co-ordinated like a living organism; and just as on seeing the fruit he will recognise the tree and the country of its growth, so from the scrutiny of the deed he can presume how it has been brought about, what has been the motives, and what kind of persons have been employed in it.[69]

What was needed, he mused in a wonderfully Collingwood-esque moment, was time for thought:

> How can one acquire the necessary precision of glance and how form the picture in any particular case? No precise rules can, indeed, be laid down on this point; but certain it is that, even in the most difficult case, if one conjure up in the mind's eye, quietly, prudently, and thoughtfully, the ways in which events have occurred, one will always arrive at a safe conclusion as to the circle or class in which persons who know something will be found.[70]

Far from limiting the process to a cerebral one, Collingwood (and Kendal) would have done well to recall that Sherlock Holmes was not averse to a bit of actual reenactment to test a theory. So too in criminal investigation proper. Based on the concept of 'context-dependent memory', physical reenactments have long been firmly established in the field of crime detection. 'In an attempt to jog the memories of possible witnesses', noted David Groome in 1999, 'crime reconstructions are often organized in which every effort is made to replicate the original events and context of the crime as exactly as possible'.[71] More recently the media was added to the mix. The first use of reenactment on television in connection with a crime was in 1976 when a police officer in Albuquerque, New Mexico, convinced a local broadcaster to make and screen a reenactment of a murder that had taken place at a service station on a busy street to encourage local residents to provide information to assist with the investigation.[72] From this humble beginning an entire genre has emerged. In 2004, it was estimated that as part of an attempt to assist the apprehension of criminals more than a 1000 programmes featuring reenactments of actual crime have been created wherever television is screened.[73]

Criminal investigators seem to share few of the hang-ups of historians. In 2003 what was then regarded as the largest reenactment in crime history was held in Victoria Park in London. In an attempt to bring to light new evidence, more than 100 people who had been in the park at the time of the murder of a student agreed to simultaneously retrace their steps while an actor re-enacted the victim's activities in their midst. Moreover it was reported that the police planned to create 'a hi-tech 3D computer picture' of the reenactment. 'The image will use graphics similar to computer games such as *Tomb Raider*', the report continued, 'allowing detectives to move around the park on a computer screen'.[74] Sherlock Holmes, if not Collingwood, would surely have approved.

This is not to suggest that the use of reenactment in criminal investigation is unproblematic. On the contrary, whatever their efficacy in the reduction of unsolved cases, some commentators have begun to raise concerns about their place in the law. Here too there is much to learn from. One problem is what is called 'unconscious transference' which arises from the use of actors. An American study has found that viewers tend either to blend features of the actor's face with those of the perpetrator or to remember the face of the actor who plays the suspect, rather than the criminal.[75] The use of increasingly sophisticated computer generated images exacerbates this problem. Similar problems have been encountered outside the law. Commentators have long meditated on the pros and cons of appropriation art, and in particular the issue of the relationship between original and copy that it raises. 'The copy at the same time restates the original *and* usurps it, repeats it *and* takes its place', writes Rex Butler, 'It both makes the originality of the original possible and destroys it'.[76] It is clear that historical reenactments have the same 'appropriation' effect, or 'unconscious transference': each successive rendition replacing important aspects of the previous one. For cinemagoers in 1970 the name Ned Kelly must have called to mind the face of Mick Jagger, as it does Heath Ledger today. Actual photos of Ned are jarringly incongruent with the familiar face on the screen. The better the mimetic realism of the reenactment, the more powerful the usurpation of the original. We must forgive viewers and jurors for believing their eyes.

More recently, reenactment has also found its way into the courtroom. In Australia the use of reenactments in court is governed by the broader rules that relate to documentary evidence. According to Arenson and Bagaric, in all jurisdictions in Australia, judges have considerable discretion in what can be admitted as exhibits, whether it is maps, models and charts or photographs and moving pictures. They also enjoy considerable latitude in relation to what are called 'views' – taking jurors to the location of an incident in order see it for themselves. The assumption that physical experience of the actual location will aid the process of understanding is worth lingering over.[77] There is a growing debate about whether what jurors see on such visits is evidence, and as early as March 2000, the Western Australian Director of Public Prosecutions aired a number of concerns at a law reform conference. Unsettled by the 'persuasive animations and graphics' evident in films such as *Jurassic Park* and *Forrest Gump*, Robert Cock QC told his colleagues that 'computer-animated crime reenactment' was fraught with 'evidentiary problems' and 'unfair prejudice'. It could look too real.[78] As Baudrillard wrote of the simulacrum: 'It is no longer a question of imitation, nor of reduplication, nor even of parody. It is rather a question of substituting signs of the real for the real'.[79]

In the US reenactment has played a larger role in courtroom proceedings, spawning an industry devoted to the production of reenactments. Take *Settlement Films* in San Diego for example. 'Although a still photograph, or a graphic display can be helpful in illustrating some aspects of the event', runs an advertisement for this company, 'nothing will explain it better than a sequence of events shown in motion':

> With videotaped reenactments you can recreate the sequence of events in the exact location where the injury has occurred. You will also be able to explore the accident scene from multiple viewpoints, demonstrate effects of obstructed sight lines, and offer your jury the perspective necessary to clearly understand the facts.

Not only is the cost of producing reenactments lower, they boast, 'The videotaped, live demonstration is also more effective, because the jury naturally relates to the dynamics of real human movements'.[80] Another company simplifies the pitch for its television program: 'No Witnesses. No Leads. No Problems'.[81]

The use of reenactment in court has also generated important case law in the US. As early as 1994, one commentator reported that several US courts of appeal had overturned convictions where the prosecution had used reenactment. For example, in *Lopez v. State* (a case described as the 'the most searching judicial analysis of the admissibility of a video-taped reenactment'), the Texas Court of Appeals found that 'any staged, re-enacted criminal acts or defensive issues involving human beings are impossible to duplicate in every minute detail are therefore inherently dangerous ... [T]he impact of reenactments is too highly prejudicial to insure the State or the defendant a fair trial'.[82]

Reflecting on the efficacy of reenactment and drawing on what was then the recent experience of working with Tony Richardson on his 1970 film, Ian Jones has described 'Dramatic reenactment of the past' as a 'tricky business'. 'Paradoxically', he commented, 'while dramatisation can utterly destroy a character – or even a whole period – it can also achieve uncanny heights of historical truth and grant us luminous glimpses of the past'.[83] Surely, like Grenville's acrid smoke, this is 'dangerous tosh'. Reenactments are becoming simultaneously more prevalent and more problematic in the law and in public education. They have become the simulacra of Baudrillard's imagination constituting what he has described as an 'obsession with historical *fidelity*' in which history itself is lost.[84] Clearly it was a mistake to lightly dismiss – as I did at the outset – the ebullient claim that Ned Kelly came to life during a reenactment in Beechworth. He did. Like it or not, historians must come to terms with these forms of 'living history': if we close our eyes they will not go away. Moreover, if Rosenstone is correct, reenactment, like film, holds out many possibilities. Without compromising on questions of rigour, the judicious historian should embrace reenactment both as a source and as a method. As sources reenactments tell us a lot about those doing the re-enacting, providing an opportunity to connect past and present by historicising the reenactments themselves. The gap between then and now is not reduced as a result, but successive reenactments provide useful sites for study along its continuum.

Despite its obvious pitfalls and dangers, there is much that a careful historian can learn about context, about material conditions, about possibility, from reenactment as a methodology. Standing on the cold stone floor metres below the 'drop' in Old Melbourne Gaol, for example, it is clear that the newspaper reporters could not have heard Ned's mumbled last words as the calico hood was pulled down over his head. People heard differently 100 years ago but not more acutely. This is evidence that should not be needlessly foregone. Although Collingwood insisted that reenactment was a cerebral activity, as a renowned archaeologist, he contradicted himself every time he held an object or shovelled some dirt. If, as E. P. Thompson argued in a well-known passage, understanding history is about getting inside episodes, reenactment must surely be pressed into service. This is nothing more than good practice for historians, summed up in R. H. Tawney's equally famous quip that historians should 'lay aside their books in favour of their boots'.[85] Walking the ground of historical enquiry, getting mud on those boots, should not be controversial. Thompson himself knew this. How did the stereotypical nineteenth-century autodidactic cobbler recite poetry while mending shoes? With a little encouragement Thompson would demonstrate.[86]

Notes

1. I. Jones (2003) *Ned Kelly: A Short Life* (South Melbourne: Lothian Books), pp. 287–8; A. Castles (2005) *Ned Kelly's Last Days* (Crows Nest: Allen & Unwin), chapter 30; *Weekly Times* (13 November 1880).
2. The Old Melbourne Gaol: Crime and Justice Experience, http://www.oldmelbournegaol.com.au/. I am grateful to Andrew Goodby of the National Trust for providing me with information about the author.
3. *Bob Hempel Presents Ned Kelly's Last Stand* (n. d.) [promotional flyer], (Benalla). See also *Sydney Morning Herald* (29 October 2001).
4. *Bob Hempel Presents*; J. Dunn (October–November 2005) 'Ned Kelly Australia's Bushranging Icon', *Outback*, 43, p. 110.
5. See *Bob Hempel Presents.*
6. Beechworth Information Centre Official Website, available at http://www.beechworthonline.com.au/visitbeechworth.asp.
7. Ibid. The nearby Burke Museum contains one of the original Ned Kelly death masks.
8. K. Kissane 'Ned Kelly Comes to Life in Beechworth', *Age*, (6 August 2007).

9. J. Cranston (2006) *The Story of the Kelly Gang* (Glenrowan: Glenrowan Cobb & Co.), p. 131; I. Bertrand and W. Routt (2007) '"The Picture that will Live Forever": The Story of the Kelly Gang', *Moving Image*, 8, p.25.
10. Jack Cranston (n.d.) 'The Story of the Kelly Gang Film', 1906–7 [press release], available at http://www.nedkellysworld.com.au/shopping/books/story_of_the_kelly_gang_film.doc; S. Jackson and G. Shirley (2006) *New Life for an Australian Classic Film* (Canberra: National Film and Sound Archive).
11. See Dunn, 'Ned Kelly', p. 110; *Border Mail*, (3 March 2007). Ned's earnings are insufficient, however, to convince the State Government to spend $200,000 to identify his remains, although this will surely change if a way to make money from the bones can be found. See *Herald-Sun* (25 May 2008). I am grateful to Marie Clarke for bringing this article to my attention. See also: P. A. Pickering (2004) 'A Grand Ossification: William Cobbett and the Commemoration of Tom Paine' in P. A. Pickering and A. Tyrrell (eds) *Contested Sites: Commemoration, Memorial and Popular Politics in Nineteenth-Century Britain* (Aldershot: Ashgate), pp. 57–80.
12. J. Dwyer and J. Schapper (eds) (2003) *Managing Our Heritage: A Review of Heritage Place Management in Victoria* (Melbourne: Heritage Council), pp. 47, 74, 93.
13. Some are at least humorously, if superficially, apposite: 'Ned Kelly Bargains (a steal)'; 'Ned Kelly's Car Rentals (for a quick getaway)'.
14. Elsewhere I have suggested that the *reductio ad absurdum* is 'traditional Aussie hot dogs' produced by KR Castlemaine in Victoria. The image on the packet of a helmeted Ned stuffing his face with processed meat product through the post-box slot rivals the bunny. See P. A. Pickering (Autumn/Winter 2008) 'The "Enormous Condescension of Posterity" Revisited: Labour Heritage and Public Memory', *Dissent*, 26, p. 44. See also I. Jones (2002) *Ned: The Exhibition: Old Melbourne Gaol 2001–2002* (Pimlico, Qld: Network Creative Services), pp. 123–4.
15. Not surprisingly numerous ballads about Ned were in circulation – and for sale – while he was still at large, but they soon went into decline after his death. Recently there has been a revival. For the soundtrack of the 1970 film the American country and western singer, Waylon Jennings, was pressed into service (as was Jagger who contributed an *a capella* version of the traditional ditty, 'Wild Colonial Boy'). In 1978, Reg Livermore, starred in *Ned Kelly: The Rock Opera*, which he had co-written with the irascible Labor party apparatchik, Bob Ellis. What this unlikely combination did for 'opera', Edouard Borovansky (1951) and Timothy Gordon (1990) have done for ballet, and Barabara Cohen (1977) Edward Cowie (1979), Graeme Koehne (2003) and Brett Dean (2007) have done for art music. Rock musicians too have mused over the Kellys. Bernard Fanning (*Powderfinger*) contributed a dirge to the soundtrack of the Jordan film and Tim Freedman (*The Whitlams*) has written a song about Kate Kelly. As lead vocalist for the Melbourne-based rock band, *Midnight Oil*, Peter Garret, now the minister responsible for the environment in the Commonwealth Government, looked improbably to a King Ned Kelly to protect the 'deadheart' of the Australian outback from American mining companies: 'if Ned Kelly was king, he'd make those robbers swing'. Another aspiring politician, John Schumann of the folk-rock band, *Redgum*, lamented the fate of 'Poor Ned Kelly' and, anticipates the

day when a 1000 'like Ned' will, more predictably, raise the republican flag.

16. E. Reade (1979) *History and Heartburn: The Saga of Australian Film 1896–1978* (Sydney: Harper & Row), p. 6.
17. *Herald-Sun* (20 April 2008).
18. Cited in *Sunday Telegraph* (23 March 2003).
19. J. M. McDonagh (2003) *Ned Kelly: The Screenplay* (Strawberry Hills, NSW: Currency Press), p. vi.
20. See Pickering, 'The "Enormous Condescension of Posterity' Revisited', p. 44. This point has also been made Brian Orndorf. See B. Orndorf (2003) 'Ned Kelly', available at www.filmfocus.com/movies/reviews.
21. C. Morris (2003) 'Ned Kelly Interview', available at www.webwombat.com.au/entertainment/movies.
22. B. McDonald (2004) *What They said about Ned!: Looking at the Legend of Ned Kelly through Books* (Bondi, NSW: Australian History Promotions), p. 26.
23. Ibid.
24. L. Rees (1973) *The Making of Australian Drama: A Historical and Critical Survey from the 1830s to the 1970s* (Sydney: Angus & Robertson), p. 48; I. Bertrand (2003) 'New Histories of the Kelly Gang: Gregor Jordan's *Ned Kelly*, available at http://www.sensesofcinema.com/contents/03/26/ned_kelly.html, p. 1.
25. R. Rosenstone (1995) *Visions of the Past: The Challenge of Film to Our Modern Idea of History* (Cambridge, MA: Harvard University Press), p. 36. Rosenstone made this point forcefully in response to a much earlier version of this chapter delivered as a paper at CalTech in 2005.
26. Rosenstone, *Visions of the Past*, chapter 9; R. Rosenstone (ed.) (1995) *Revisioning History: Film and the Construction of a New Past* (Princeton: Princeton University Press), p. 202.
27. See D. Stewart (1963) *Ned Kelly* in *Three Australian Plays* (Harmondsworth: Penguin Books), pp. 97–236; A. A. Phillips (September 1956) 'Douglas Stewart's *Ned Kelly* and Australian Romanticism', *Meanjin*, pp. 261–71. In opposition to Kelly and Joe Byrne, the Ulysseans, Phillips identifies Edwin Living, a teller in the Bank of New South Wales, and Thomas Curnow, the teacher who betrayed Ned, as Telemachians. Kelly has also been the subject of numerous novels which are not considered here. See McDonald, *What They said about Ned!* for a comprehensive list. See also G. Huggan (May 2002) 'Cultural Memory in Postcolonial Fiction: The Uses and Abuses of Ned Kelly', *Australian Literary Studies*, 20, pp. 142–54.
28. It is notable that its treatment in recent literature, notably Peter Carey's outrageously titled True History of the Kelly Gang, has been postcolonial in genre. See P. Carey (2000) *True History of the Kelly Gang*, St Lucia, Qld: University of Queensland Press. See Huggan, 'Cultural Memory in Postcolonial Fiction', for a most interesting discussion of Carey and Robert Drewe's, *Our Sunshine*. See R. Drewe (1991) *Our Sunshine* (Sydney: Pan Macmillan).
29. The film failed at the box office although this was attributed by some to simple fact that Kelly was virtually unknown outside of Australia. See *Herald-Sun* (20 April 2008).
30. See K. C. Barton and L. Levstik (2004) *Teaching History for the Common Good* (London: Lawrence Erlbaum Associates), pp. 119–21; R. Rosenzweig and

D. Thelan (1998) *The Presence of the Past: Popular Uses of History* (New York, Columbia University Press), pp. 105–6.

31. R. Pearce (August 2003) 'University History', *History Today*, 53, p. 54; *The Times* (21 July 2003).
32. Greg Wong (14 November 2006) 'Educators Explore "Second life" Online', CNN. I am grateful to Kate Bowan for this reference.
33. There have been other problems casting Ned. In Rupert Kathner's 1951 film, *The Glenrowan Affair*, Ned was played by Bob Chitty, then a leading footballer reputed to be one of the toughest exponents of the Australian national game during the years when its brutality was taken for granted. Although a controversial choice Chitty was certainly tough enough: in 1943 he had turned in a best-on-ground performance for his club Carlton on Saturday despite having lost the top part of a finger in an industrial accident during the week. See J. Ross (ed.) (1996) *100 Years of Australian Rules Football* (Ringwood: Viking), p. 169.
34. See I. Jones (1973) 'Re-Enacting the Past' in D. Duffy, G. Harman and K. Swan (eds) *Historians at Work* (Sydney: Hicks Smith & Sons), p. 259; J. Baudrillard (1995) *Simulacra and Simulation* (Ann Arbor: University of Michigan Press).
35. The press reported that he said: 'Well, I expect to go a little further on, and to see you there'. See *Weekly Times* (30 October 1880).
36. Tony Richardson died in 1991, and unfortunately I have been unable to contact Ian Jones to ask if he recalls the reason for the different cuts.
37. I. Bertrand (1978) *Film Censorship in Australia* (St. Lucia, Qld: University of Queensland Press), p. 12. Bertrand argues that as theatre came to be seen as an elite recreation, it increasingly escaped close scrutiny by the censor. Nevertheless Stewart's play was banned in Melbourne in 1956 as it was felt that it did not project an appropriate image during the Olympics. The irony that the Opening Ceremony of the 2000 Sydney Games featured several hundred armed Neds has not been noted, but it is further evidence of the way in which the outlaw's image has been commodified into meaningless abstraction in the intervening years.
38. I. Bertrand, *Film Censorship in Australia*, pp. 1–2.
39. Rees, *The Making of Australian Drama*, pp. 8, 87–91.
40. E. J. Hobsbawm (1972) *Bandits* (Harmondsworth: Penguin Books), p. 131. The leading exponent of this view is John McQuilton (1979) *The Kelly Outbreak 1878–1880* (Melbourne: Melbourne University Press).
41. For his reference to Kelly, see Hobsbawm, *Bandits*, pp. 17and 131.
42. Cited in Phillips, *Trial of Ned Kelly*, p. 98.
43. *Bulletin* (3 and 24 January 1907).
44. Bertrand, *Film Censorship*, pp. 42, 97, 108, 110–11, 119, 123. According to Lloyd Hughes 'bushranger movies' accounted for 13 per cent of all Australian films made in the 1910s. See L. Hughes (2005) *The Rough Guide to Gangster Movies* (London: Rough Guides), p. 248. I am grateful to Daniel Pickering for this reference.
45. *Bulletin* (2 May 1907).
46. One of the few to oppose the censorship was the labour activist and journalist, Henry Boote. Writing in 1921 in Sydney, Boote wished 'confusion to the censors who would muzzle our mouths and blindfold our eyes' and declared himself in favour of 'an absolutely free theatre, just as I am in favour of an

absolutely free platform, and an absolutely free press'. Cited in Bertrand, *Film Censorship*, p. 18.

47. Cited in Bertrand, *Film Censorship*, p. 165.
48. See Bertrand, *Film Censorship*, pp. 110–11, 123, 165, 210.
49. See B. Caleo (August-September 2000) 'Such is Myth: Ned Kelly Reports from the Dock', *Arena Magazine*, p. 25; J. Burnside, 'Regina v Edward (Ned) Kelly', available at http://www.ironoutlaw.com/html/writings_burnside_01.html; (22 August 2000) 'Ned Kelly Retrial: Wilful Murder or Self-Defence' [transcript], *The Law Report*, ABC Radio National.
50. See C. Strange (2006) 'Hybrid History and the Retrial of the Painful Past', *Crime, Media, Culture*, 2, pp. 197–215. Ray Surette and Charles Otto have argued that television has limited learning impact. See C. Otto and R. Surette (2002) 'A Test of a Crime and Infotainment Measure', *Journal of Criminal Justice*, 30, p. 445.
51. And it predated the well-known 're-trial' of an infamous rebel in Canada. Thus Strange is incorrect when she argues that the trial of Riel was a first. See Strange, 'Hybrid History', p. 199.
52. I. Jones (1977) 'Introduction' in R. Simpson *Trial of Ned Kelly* (Ringwood: Heinemann Educational Australia), p. viii.
53. Burnside's analysis of the forensic evidence casts doubt on Kelly's version of events as does Alex Castles's. See Burnside, '*Regina v. Edward* (Ned) Kelly'; Castles, *Ned Kelly's Last Days*, pp. 32–8, 163. See also Phillips, *The Trial of Ned Kelly*, chapter 14. I am grateful to Simon Bronitt for comments on Kelly's trial. We are planning to convene a conference on this subject in 2009. Kelly's own account of events, in his famous Jerilderie letter, was suppressed by the action of his own Counsel, but as Castles notes, it does not alter the facts of the episode. See E. Kelly (2001 [1897]) *The Jerilderie Letter* (London: Faber).
54. Jones, *Ned Kelly*, p. 301.
55. The best construction that Jones can put on it is that it was 'a bloody, ugly, unglamorous gun brawl with good, honest policemen dying before the guns of a good honest rebel'. See Jones, 'Re-Enacting the Past', p. 257.
56. Directed by Mark Lewis, it was produced by Windfall Films and Look Television for the Australian Broadcasting Corporation in association with Channel 4 and National Geographic Channels International. I am grateful to Simon Bronitt for bringing this to my attention and Will Davies from Look Television for providing me with a copy of the documentary. Another documentary to frame the story in extreme language (was Kelly 'an outlaw, a freedom fighter or a murdering terrorist') and to interview descendents is 'Besieged – The Ned Kelly Story', screened on a commercial network in December 2003. It was written and directed as a collaborative project between Australia and Ireland by Barrie Dowdall and Gregory Miller. Despite the promise of 'richly layered dramatizations re-enacting scenes from the story and readings from original texts (such as the Jerilderie letter), location filming, expert commentary from leading Kelly historians', the documentary offers an unexceptional and basically sympathetic account. The film was financed by the Regional Victoria Film location Assistance Fund, the Australian Film Commission and the Irish Film Board.
57. See 'In the Matter of a Bail Application by Steven Wayne Hillier [2003] ACTSC 50 (25 June 2003)', available at http://www.courts.act.gov.au/

supreme/judgments/hillier.htm. See also K. J. Crispin (n.d.) 'Of Augeries and Experts', available at http://www.law.monash.edu.au/conference/expertconf/crispin.pdf. I am grateful to Simon Bronitt for bringing these referecnes to my attention.

58. See R. Wheate (2007) 'Jury Comprehension and the Use of Forensic Evidence' (Ph.D. dissertation, University of New South Wales at ADFA).
59. (2006) 'Jack the Ripper: The First Serial Killer', Discovery Channel, screened on ABC television, 7 August 2008.
60. I am grateful to Fred Inglis for the latter description.
61. G. Dening (1992) *Mr Bligh's Bad Language* (Cambridge: Cambridge University Press), pp. 4–5. Dening wrote: 'I am not much for re-enactments. Re-enactments tend to hallucinate a past as merely the present in funny dress. They give modernity and fashion a fillip by making the past look quaint. They patronize the human condition in hindsighted superiority. They remove the responsibility of remedying the present by abstracted, unreflective search for details whose remedy will make no difference'.
62. See J. Sullivan 'Making a Fiction of History', *Age* (21 October 2006).
63. I. Clendinnen (2006) 'The History Question. Who Owns the Past?', *Quarterly Essay*, 23, p 20, 22.
64. K. Grenville (2006) *Searching for the Secret River* (Melbourne: Text Publishing Company), p. 47.
65. K. Grenville, *Searching for the Secret River*, p. 161; Clendinnen, 'The History Question', p. 24.
66. R. G. Collingwood (1971 [1946]) *The Idea of History* (Oxford: Oxford University Press), p. 282.
67. Collingwood, *The Idea of History*, p. 243.
68. I am grateful to Fred Inglis for pointing out Collingwood's passion for detective fiction. He is unsure if Collingwood also studied textbooks on criminal investigation.
69. N. Kendal (1934) *Criminal Investigation: A Practical Textbook for Magistrates, Police Officers and Lawyers* (London: Sweet & Maxwell), p. 37. The first edition by J. Adam and J. Collye Adam was published in 1906.
70. Kendal, *Criminal Investigation*, p. 37. The other pioneer of criminal investigation techniques, Edmond Locard, made a direct link between the criminologist and the archeologist: 'the criminal re-creates the criminal from the traces the latter leaves behind, just as the archeologist reconstructs prehistoric beings from his finds'. Cited in W. J. Chisum (January 2000) 'Evidence Dynamics: Locard's Exchange Principle and Crime Reconstruction', *Journal of Behavioral Profiling*, 1, p. 3.
71. D. Groome (1999) *An Introduction to Cognitive Psychology: Processes and Disorders* (Hove: Routledge), p. 120.
72. D. Challinger (2004) *Crime Stoppers Victoria: An Evaluation* (Canberra: Australian Institute of Criminology, available at http://www.aic.gov.au/publications/tbp/tbp008/, p. 6. The officer was Greg MacAleese.
73. Ibid.
74. *Daily Telegraph* (3 March 2003).
75. R. Roesch, R. Corrodo and R. Dempster (eds) (2001) *Psychology in the Courts: International Advances in Knowledge* (London: Routledge), pp. 242–3.

76. R. Butler (ed.) (1996) *What is Appropriation?: An Anthology of Critical Writings on Australian Art in the '80s and '90s* (Sydney: Power Publications and IMA), p. 15.
77. K. J. Arenson and M. Bagaric (2005) *Rules of Evidence in Australia: Text and Cases* (Sydney: LexisNexis Butterworths), pp. 383–94.
78. R. Cock (2003) 'Evidentiary Rules and Aids in the Presentation of Evidence', *Reform of Criminal Trial Proceedure*, conference papers, paper 3, available at http://www.aija.org.au/ctr/cock.htm, pp. 3–4.
79. Baudrillard, *Simulacra and Simulation*, p. 2. Baudrillard gives sustained attention to the relationship between history and cinema. See his chapter 'History: A Retro Scenario' in *Simulacra and Simulation*, pp. 43–51.
80. See Legal Re-Enactments, available at http://www.settlementfilms.com/legal/re-enactments.html.
81. See 'Forensic Files: No Witnesses. No Leads. No Problem.', available at http://www.forensicfiles.com/about-forensicfiles.htm. On a smaller scale pioneering companies in Britain and Australia will conduct historical re-enactments to order. See for example: Chance Encounters, available at http://www.chanceencounters.co.uk/company.html and Historica: Bringing History to Life, available at http://www.historica.com.au/.
82. D. B. Hennes (1993) 'Manufacturing Evidence for Trial: The Prejudicial Implications of Videotaped Crime Scene Re-enactments', *University of Pennsylvania Law Review*, 142, pp. 2142–3.
83. Jones, 'Re-enacting the Past', p. 251.
84. Baudrillard, *Simulacra and Simulation*, p. 47.
85. Cited in E. Kerridge (1951) 'Ridge and Furrow and Agrarian History', *Economic History Review*, New Series, Vol. 4, No. 1, p. 14 and 14n.
86. E. P. Thompson (1979 [1965]) 'The Peculiarities of the English' in *The Poverty of Theory and Other Essays* (London: Merlin Press), p. 65. I am grateful to Stephen Roberts for the story of Edward re-enacting an autodidact.

9

R. G. Collingwood, Historical Reenactment and the Early Music Revival

Kate Bowan

As a young boy, Robin George Collingwood would steal from his bed and sit on the stairs with his sisters shrouded in darkness secretly listening to his mother playing the piano below. In this way, he remembered in his autobiography, he became familiar with all the Beethoven sonatas and most of Chopin's music.[1] Music, it turns out, played not only an important role in Collingwood's personal life, but was also central to his preliminary formulations of historical reenactment. The circles in which his father moved brought him into contact with what came to be known as the Early Music Revival pioneered in England by the eccentric French émigré, Arnold Dolmetsch. The influence of this movement, which centres on a historically considered reconstruction of music from a distant past, can be detected in his early writings on reenactment.

To perform music is to bring it to life. Every sounding realization of a written score is inescapably a reenactment. The relation of musical performance to history and music's own historicity is however a complex issue. Some approaches to performance are more consciously historically informed than others and can be seen as a kind of sonic historical reenactment. I will limit my focus to this particular field of musical performance known as early music or historical performance; a style of performance that strives for authenticity; that tries to faithfully adhere to the composer's intentions and reproduce the work as it would have sounded in the composer's own time on period instruments using historically correct techniques. In this way it reflects many of the ideals of historical reenactment. As an approach to past music, historical performance has gained enormous momentum throughout the last century and is today one of the liveliest fields of European art music production with far-reaching influence. It is, as Frederick Neumann has pointed out, 'a social and cultural phenomenon of great complexity'.[2]

It is perhaps strange that European art music, which has devoted itself almost exclusively in the last hundred-odd years to the performance of music from the past, creating what philosopher Lydia Goehr has famously called an 'imaginary museum' of musical works,[3] should also have produced a subculture that also performs old music but which, nonetheless, sees itself as fundamentally different to the mainstream tradition. Mainstream performance has grown out of the nineteenth-century tradition of the public concert, the symphony orchestra and the idea of the virtuoso. Virtuosity, technical precision, and professionalism are central to its culture. It has constantly transformed itself and has long performed past music upon modern instruments according to contemporary and continually evolving performing conventions. The early music revival, which began in earnest towards the end of the nineteenth century, was a reaction against this mode of performance. This now-diverse field has since sought not only to recover lost repertories, instruments and practices, but also to rescue earlier repertoire from what it considered the mainstream's ahistorical treatment. It has its roots in period instrument reconstruction, historically correct playing practices (which includes a consideration of the spatial dimensions of the original venues) and the scholarly study of a wide range of textual and iconographical sources. It was born from an amateur performance tradition that has however, over the last 25 years or more, been professionalized to a similar level as that of the mainstream. It has developed into a highly specialized musical culture that has produced a large and ever increasing body of literature. Its practitioners include, as outlined by gambist/musicologist Laurence Dreyfus, performers (professional and amateur), teachers and scholars (mostly from musicology but also from the fields of dance, theatre, iconography and cultural history), instrument makers, publishers, critics, concert managers, agents, recording companies, sound engineers and audiences.[4]

Although it originally focussed on seventeenth- and eighteenth-century music, the historical performance movement's attention has moved in both temporal directions. Now Brahms's symphonies written in the last decades of the nineteenth century are performed on original instruments and the 'lost traditions' of the middle ages and earlier have been reconstructed or, as some would suggest, reinvented. As such it has proven to be an innovative and enormously influential phenomenon in the world of art music.

In the quest for the 'authentic' reconstruction of the past down to its finest detail, the devotees – both scholars and performers – of early music have seemingly anticipated the current interest in reenactment by several decades. The vigorous and at times fiercely acrimonious debates surrounding key issues that have dominated this musical scene, particularly the long-standing debate over authenticity, bear many resemblances to certain of those that interest academic historians detained by the possibilities of reenactment nowadays.

I am not a specialist in historical performance (and they are plentiful); the aim is to point out to non-musicians the important contributions this lively and highly controversial musical field can make to ideas of reenactment, and to highlight the role music played in Collingwood's formulation of the idea of historical reenactment.

The Collingwood home was filled with music. His mother was a fine pianist and was Robin's first teacher; his father had a fine singing voice. In addition to drawing, painting and mapmaking, Collingwood, by the time he was 12, had also learned to play the piano, to sing, and to compose.[5] He remembered many evenings 'spent playing and hearing music'.[6] In 1903 when he was sent to Rugby, which he hated, music became a source of escape. Here he 'discovered Bach, learned to play the violin, studied harmony and counterpoint and orchestration, and composed'. His aptitude was recognized with a prize for music theory and composition.[7] His study of music continued at Oxford, where he took vocal lessons while a tutor, all the while producing 'enormous quantities of songs and instrumental pieces', which he later dismissed as 'trash'. The idea of becoming a full-time composer persisted into his young adulthood and the violin was only discarded after his first marriage.[8] He was therefore intimately acquainted with music both as a composer and performer.

In 1928, while on vacation in France, at Le Martouret, 'that pleasant country-house near Die', Collingwood sat down under the plane trees on the terrace and mused on the nature of musical performance and its relation to the study of music history. This train of thought provided the inspiration for his preliminary theories of historical reenactment which were ultimately presented as the essay, 'Outlines of a Philosophy of History'.[9] They took the shape of three propositions: history as 'the history of thought'; the related idea that 'historical knowledge is the reenactment in the historian's mind of the thought whose history he is studying', and thirdly 'historical knowledge is the reenactment of a past thought incapsulated [*sic*] in a context of present thoughts which, contradicting it, confine it to a plane different from theirs'.[10] Although music did not make it into 'History as Reenactment of Past Experience' in the *The Idea of History*, it looms large in the 1928 essay:

> No historian of music deserves the name unless he has studied for himself the old music whose growth and development he is trying to describe. He must have listened to Bach and Mozart, Palestrina and Lasso, and possess personal acquaintance with their works. This means that he must have been present at actual performances of these works either physically or in imagination; and in the latter case the imaginative power is acquired only by actually hearing similar things performed – e.g. a man who had never heard an orchestra of the Beethoven period could not read a symphony of Beethoven in score with any chance of obtaining a good imaginative hearing of it. We may therefore boldly say that the *sine qua non* of writing the history of past music is to have this music *re-enacted in the present.*[11]

What is then crucial to historical understanding, its '*sine qua non*', is the perception of the object, in this case the *listening to* of a piece of music. This is a key contradiction to Collingwood's later thinking, namely that in order for the historian to re-enact the past *in the mind* a somatic experience is needed. His assertion that one could *not acquire* 'a good imaginative hearing' of a Beethoven symphony without having *heard* an orchestra from Beethoven's time shows that performative experience was central to his initial conception of historical reenactment. This contradiction in his early theorizing – this need for performance – gives strength to the case for actual, embodied reenactment itself. William Dray has noted the important role musical performance plays in this early essay, but he quickly dismisses it, focussing instead on the idea of re-enacting the creative process.[12] Collingwood, however, was not trying to understand or re-enact the compositional process; for him performance brought the past into the present as lived experience.

Although Collingwood fondly remembered his mother playing Beethoven and Chopin, he hastened to assure readers of his autobiography that they were his mother's favourite composers, not his. Later, when sitting the entry examinations for Oxford, in a gesture of defiant non-conformity he decided to behave 'disgracefully' and devoted his 'general' paper to a discussion of Turner and Mozart. The distancing from Beethoven and Chopin, the decision to write about his beloved Mozart and the other composers he used as examples in the 1928 essay, namely Bach, Palestrina and Lasso, betrays a refined and restrained musical sensibility; more Apollonian than Dionysian, one that was ill-disposed to the emotional excesses of the late romantic sensibility. His inclusion of the pre-1600 composers, Palestrina and Lasso particularly, reveals a familiarity with early music that was at that time in no way typical. Collingwood, as seen above, was musically well educated; his awareness of these composers who figured so prominently in the Early Music Revival is perhaps not extraordinary, but there was a personal connection that may have further exposed him to what was then a fringe artistic movement. This connection came through his father.

The relationship between father and son was close. William Gershom Collingwood, artist, author and antiquary, had educated his son until his early teens and they continued to collaborate on archaeological ventures until William's death in 1927. His influence on his son was profound. After meeting John Ruskin at Oxford, William was enlisted as his assistant, a role that he held for many years. It is through this close friendship with Ruskin that he met the founder of the Art and Crafts Movement, William Morris. And he is remembered as having moved in William Morris's circle.[13] Given that R. G. Collingwood spent long intensive years in the company of his father, he too would have participated, albeit as a child, in this milieu. Notably William Morris, towards the end of the nineteenth century, had become aware of, and deeply interested in, the work of the pioneer of the Early Music Revival, Arnold Dolmetsch.

Dolmetsch was a French-born violinist who was to become not only a performer, but also an instrument maker, scholar, teacher and entrepreneur. He came from a family of instrument makers and had trained in the family factory, Maison Dolmetsch-Guillouard, which produced pianos, organs and harmoniums. After studying violin and composition at the Brussels Conservatoire he came to England to study at the newly founded Royal College of Music (RCM) in 1883. He quickly turned to the study of early English instrumental music. The Early Music Revival was emerging in many European centres; in England it was swept up into the broader nationalist revivals of music, literature and drama, including not only the Elizabethan Revival but also the Folk and Celtic Revivals. Nor was Dolmetsch the only musician in England to be excavating the past; he took his part alongside others such as Richard Terry, Francis Galpin and A. J. Hipkins. But Dolmetsch, more than the others, lived his life as a historical reenactment in the current sense of the idea. He was so 'enamoured with ancient music,'[14] so determined to bring ancient music back to life, he immersed himself and his family in a past that came alive for him as embodied musical experience.

In 1889, four years after the showing of 'obsolete instruments' at the International Inventions Exhibition in South Kensington, Dolmetsch accidentally bought a viola d'amore at an auction, mistaking it for a modern viola. He went on to restore it and then began to collect other antique instruments. Soon restoration expanded into reproduction. He also began to trawl the British Museum and RCM library collections unearthing countless musical scores, working tirelessly towards the restoration and recreation of past music and its instruments. This passion consumed the rest of his life. He collected and built viols and lutes, psalteries, virginals, clavichords and harpsichords and later on fortepianos, harps, rebecs, Baroque violins and vihuelas not to mentions the recorder (for which he is best-known), which he began to produce in the 1920s. He had found his calling and in his approach laid the foundations for the early music movement. 'This music', he proclaimed, 'is of absolute and not antiquarian importance; it must be played as the composer intended and on the instruments for which it was written with their correct technique'.[15] In 1915 he wrote *The Interpretation of the Music of the Seventeenth and Eighteenth Centuries*, the result of his study of early performing styles and notation through the treatises and manuscripts.[16] In 1925 he started annual festivals of early music at his home at Haslemere and in 1928, the same year that Collingwood was pondering reenactment, Dolmetsch founded his own eponymous foundation 'for the encouragement of the revival of early instrumental music'.[17]

His first public concert featuring period instruments in June 1890 featured many little-known English composers. He promoted an informal mode of musical performance quite different to that of the nineteenth-century public concert. Although staging a number of public concerts during the 1890s at Barnard's Inn and later Clifford's Inn in Holborn and later at the 'Fitzroy Settlement', by 1893 he opted for the intimacy provided by performing with his family and friends at their family home in Dulwich. It was not surprising that Dolmetsch named the house 'Dowland' after the Elizabethan composer John Dowland. The first concert series included works by composers that had 'not in living memory been heard upon the instruments for which they were written' and which were to be performed 'according to the original intentions of the writer'.[18]

Dolmetsch's concerts quickly gained the attention of George Bernard Shaw who, apart from his other occupations, was a discerning and prolific music critic. Shaw found Dolmetsch's viol concerts both historically and musically enjoyable. He believed that the revival of early English music was vital to the rebuilding of an English national music and was convinced that:

> In this revival the old music must serve as a starting point, just as XIII century work has served, and is serving, in modern revivals of the

> other arts. That is why I attach such importance to these concerts of Mr Dolmetsch, which are, besides, highly enjoyable both to experts in music and to the ordinary Englishman.[19]

He articulated the connection between the objectives of the French-man and William Morris in his account of Dolmetsch's first reconstruction of a clavichord. It was, he surmized, 'likely to begin such a revolution in domestic instruments as William Morris's work made in domestic furniture and decoration, or Philip Webb's in domestic architecture'.[20] Shaw also realized the importance of authenticity to Dolmetsch's work: 'Now, for some time past Mr Arnold Dolmetsch has been bringing the old instrumental music to actual performance under conditions as closely as possible resembling those contemplated by the composers'.[21] His prediction that 'the movement will have the capacity for becoming a popular movement' was prophetic.[22]

Dolmetsch's interest in the reconstruction of instruments and music from the past resonated with the interest in medieval and Renaissance culture of many in the Arts and Crafts Movement. Morris had been taken to a Dolmetsch concert in 1893 and had, according to Dolmetsch, instantly understood the music. Morris, Dolmetsch asserted, 'had found the lost art' and was moved to tears.[23] It was the first of many concerts he attended, becoming a regular visitor to 'Dowland'. Shaw did not miss the significance of Morris's interest. For Morris, Shaw maintained, hated the piano and would 'not go to an ordinary concert', but was sorely tempted, Shaw further reported, 'to try his hand at making fiddles; and has been seen at one of Dolmetsch's viol concerts apparently enjoying himself'.[24] Dolmetsch's first harpsichord constructed at the behest of William Morris was completed in 1896 for the Arts and Crafts Exhibition Society. Dolmetsch was one of Morris's many friends who tried to comfort him as he was dying. Fiona McCarthy describes a visit of 21 September 1896 when he came bringing his own instrument and played Tudor virginal music to his dying friend, including a pavane and galliard by the Elizabethan recusant composer, William Byrd. Morris 'enjoyed it extremely, but was a good deal affected'.[25]

It was not only the Shaw and Morris families who attended Dolmetsch concerts (Shaw's sister, Lucy Carr Shaw, sang at some concerts). Dolmetsch concerts provided a meeting place for the London literati and attendant bohemian set. The Shaws and Morrises were rubbing shoulders with Yeats, Joyce, Symons, Roger Fry,[26] the Pounds and many others. Dolmetsch's quest to recreate ancient music captured their imaginations. Despite standing not quite five feet tall, he was a powerfully charismatic and 'mercurial' personality.[27] Whereas many in the musical establishment regarded him as somewhat of a bohemian crank (although he did have influential supporters), the Bloomsbury set embraced him. The former perhaps looked askance at his penchant to dress in pre-Raphaelite garb (such as velvet suits, knee britches, lace ruffles and shiny shoe-buckles), the latter loved him for it. 'Owning a Dolmetsch instrument', Harry Haskell observes in his history of the Early Music Revival, 'became a status symbol in smart society'.[28] Yeats bought a psalter for his friend, Florence Farr, to play while reciting his poetry. The Pounds received a clavichord from Dolmetsch as wedding gift. It became a prized possession that Ezra would drag around Europe with him.[29] Joyce 'approved' of Pound's clavichord, but his own endeavours to acquire a lute so as to 'coast the South of England from Falmouth to Margate singing old English songs' was thwarted by Dolmetsch, who simply refused to make him one. Wyndham Lewis was less positive about Dolmetsch's gift and Pound suspected he saw it as 'a strange unaccountable sort of mouse-trap'.[30] Although considered marginal by the musical establishment, Dolmetsch achieved a certain notoriety. His performances brought the likes of Isadora Duncan and Loie Fuller to their feet. Ezra Pound, an admirer and a self-confessed 'crank', wrote approvingly of Dolmetsch's precise scholarship in an essay of 1918, in which he also noted dryly that 'They tell me "everyone knows Dolmetsch who knows of old music", but not many people know of it'.[31] Collingwood then, as revealed in his 1928 essay, was one of the minority.

Dolmetsch proved to be a captivatingly eccentric muse for many of these writers. He featured in several literary works written by many attendees of his concerts including Ezra Pound, Arthur Symons, James Joyce and George Moore. He appears in Pound's *Canto LXXXI* after the musical subtitle 'libretto':

> Yet
> Ere the season died a-cold
> Borne upon a zephyr's shoulder
> I rose through the aureate sky
> *Lawes and Jenkyns guard thy rest*
> *Dolmetsch ever be thy guest,*
>
> Has he tempered the viol's wood
> To enforce both the grave and the acute?

> Has he curved us the bowl of the lute?
> *Lawes and Jenkyns guard thy rest*
> *Dolmetsch ever be thy guest,*[32]

Arthur Symons identifies the nostalgia – 'the melancholy desire of ancient things' – that drove Dolmetsch's vision (the same one that fed the Arts and Crafts Movement), in his poem dedicated to Dolmetsch, 'On an Air of Rameau'.[33] Dolmetsch was immortalized in *Ulysses*, his name appearing in a conversation between Bloom and Stephen Daedulus on Elizabethan music:

> Stephen, in reply to a politely put query, said he didn't sing it but launched out into praises of Shakespeare's songs, at least of in or about that period, the lutenist Dowland who lived in Fetter lane near Gerard the herbalist, who ANNO LUDENDO HAUSI, DOULANDUS, an instrument he was contemplating purchasing from Mr Arnold Dolmetsch, whom B. did not quite recall though the name certainly sounded familiar, for sixty-five guineas and Farnaby and son with their DUX and COMES conceits and Byrd (William) who played the virginals, he said, in the Queen's chapel or anywhere else he found them and one Tomkins who made toys or airs and John Bull.[34]

The lesser known Gaelic Revivalist, George Moore, was so taken with Dolmetsch's restorative mission, he based the protagonist's father on him in a kind of literary reenactment with his novel, *Evelyn Innes*. The story revolves around Evelyn's father, Mr Innes, who collects early instruments which he restores and plays at his home, Dowland.[35] Moore sought a realism in his writing and indulged in almost superfluous detail, as well as soliciting experts to write certain sections of his fiction, one of whom was Dolmetsch himself. Note the breadth of knowledge of pre-seventeenth-century composers in the following excerpt:

> For Mr. Innes's ambition was to restore the liturgical chants of the early centuries, from John Ockeghem, the Flemish silver-smith of Louis XI, whose recreation it was to compose motets, to Thomas da Vittoria; and, after having made known the works of Palestrina and of those who gravitated around the great Roman composer, he hoped to *disinter* the masses of Orlando di Lasso, of Goudimel and Josquin des Près, the motets of Nannini, of Felice Anerio, of Clemens non Papa ... He would go still further back.

Dolmetsch's desire to salvage a pure authentic music looms large over the novel, seen particularly in this discussion of lost traditions, vulgar nineteenth-century bowdlerization, and the vexed question of translating ancient notation into modern:

> For before this music was the plain chant or Gregorian, bequeathed to us by the early Church, coming down to her, perhaps, from Egyptian civilisation, the mother of all art and all religion, *an incomparable treasure which unworthy inheritors have mutilated for centuries*.
>
> It was Mr. Innes's belief that the supple, free melody of the Gregorian was lost in the shouting of operatic tenors and organ accompaniments. The tradition of its true interpretation had been lost, and the text itself, but by long study of ancient missals, Mr. Innes had penetrated the secret of the ancient notation, vague as the eyeballs of the blind, and in the absence of a choir that could read this strange alphabet of sound, he cherished a plan for an edition of these old chants, re-written by him into the ordinary notation of our day.[36]

Like Shaw, Moore also predicted the fervent zeal with which many would take up early music. He described Mr Innes's students as 'disciples', who would help 'to preach the new musical gospel'.

The auditory sense was not the only one Dolmetsch sought to excite. The visual dimension was also key to Dolmetsch's mission (Figure 9.1). The critic for the *Westminster Gazette* described attending a concert at Barnard's Inn as 'a sudden plunge from the 19th to the 17th century'.[37] Those who attended a concert at 'Dowland' were subjected to an immersive experience. 'Dowland', a two-storey yellow brick villa, had itself been radically altered, particularly the music room. His biographer, Margaret Campbell offers this description:

> He placed his beloved little 16th-century chamber organ in the centre of the raised platform extending from the front bay window, and stood the keyboard instruments round the sides of the room; the walls were hung with viols, violins and lutes, whilst high backed chairs and candles in brass sconces *completed the picture of the past come to life*.[38]

Thus, the audience entered a quasi-Elizabethan world; a calm, otherworldly place – a refuge from modern life. Mabel, Dolmetsch's third wife, paints a similarly vivid picture of a house concert at 'Dowland':

> The concert room, tinted a soft diaphanous green, was entirely illuminated by wax candles, set round the walls in hand-beaten brass sconces, and interspersed with rare lutes and viols, suspended from hooks.[39]

This re-enactive enterprise was not restricted to his home. An evocative description of a concert at the home of Violet Gordon Woodhouse, a pianist who turned to the harpsichord with great success after meeting Dolmetsch, bears a marked similarity to Mabel's account. Violet had transformed the library into a 'retiring' chamber by letting down a 'thick green velvet curtain' through which the Dolmetsch family replete in Elizabethan costume emerged to take their places at their instruments waiting for them on the raised dais. In many ways the setting closely resembled Dowland's:

> There being no electricity in the house, the drawing-room was lit by two statues holding oil lamps in the niches on either side of the double

Figure 9.1 Dolmetsch family, Jesses, Haslemere, 31 August 1928

> doors, and by violet-coloured wax candles set on the green-striped walls in flat brass sconces. Since no paintings could compete with the beautiful ceiling, Violet had hung the room with silhouettes on glass and engravings in low tones of brown, grey and black. The dark, polished boards of the uncarpeted floor gleamed in the candlelight, and high piles of cushions haphazardly filled the window seats.[40]

This inclination towards historical reenactment and the overriding concern with authenticity made Dolmetsch attractive to like-minded individuals such as William Poel, the founder of the Elizabethan Stage Society, with whom he was to collaborate. He first worked with Poel in 1895, a year after the society had begun. Both rejected the late nineteenth-century aesthetic and shared the objective to produce work that recreated as faithfully as possible the author's intentions and the time from which it came. Poel developed an Elizabethan stage that produced what has been described as 'the first historically authentic productions' of Shakespeare.[41] It was an approach that has had a long-lasting influence on the Globe Theatre. Dolmetsch's brief was to provide historically suitable music to accompany a range of Shakespeare's plays. The Dolmetsch family obliged, providing historically appropriate performances on lutes, virginals, viols, viols de gamba and the pipe and tabor, dressed again in period costume.

From its beginning the early music movement has been obsessed with questions of authenticity and historical accuracy. This was due in large part to Dolmetsch who has been generally understood to be the seminal figure of the Early Music Revival; 'the true founder of the authenticity movement'.[42] 'It was' Haskell sums up, 'he who set the agenda for and defined the issues addressed by the revival. He anticipated and articulated the preoccupations of latter-day early musicians, notably their concern for "authenticity"'.[43] The notion of authenticity was for a long time based on the concept of *Werktreue* – an absolute fidelity to the work and the composer's intentions: to play it as it was played in the time in which it was written.[44] As Haskell pointed out: 'Authenticity is, of course, the nub, the central issue, the very *raison d'etre* of the early music movement. In a sense, the history of the movement is the history of the search for authenticity – or more accurately, the history of changing concepts of authenticity – in the performance of early music'.[45]

Being authentic allows the possibility of being inauthentic. In the realm of artistic performance which has aesthetic as well as historical value this distinction can become fraught. To say the historical performance movement is a highly contested site is an understatement.

It has long been a battleground on which passions have raged and boiled. The debates over how past music have been characterized in bellicose terms such as 'turf wars' and the 'authenticity wars'. The conflict has played out along lines of fiercely held and ideologically driven beliefs. Accusations of pedantry, dogmatism, dullness, inflexibility, amateurism, puritanism, antiquarianism, fetishism, childishness, ahistoricism and inauthenticity have been thrown to and fro. In 1992, Joseph Kerman noted that Early Music had 'always flourished in an atmosphere of multiple controversy, ranging from a philosophical argument about music and history to a professional dispute over turf'.[46] As recently as 2004, editor of *Early Music*, Nicholas Kenyon, referred to the debate which had, in his estimation, 'become excessively tortured'.[47]

Historical performance has produced a body of literature whose concerns and the debates in some respects bear an almost uncanny resemblance to those of historical reenactment and its developing intersection with historical scholarship. As a result a large and rich body of literature now exists. However a musical performance, even if historically informed, is also a work of art for which questions of musical and aesthetic value are as important as historical verisimilitude. Like a historical reenactment, a musical performance gives the past a vivid immediacy and moreover it can be repeated over and over again. As such it provides food for thought on a variety of related concerns including authenticity, as well as questions of affect and historical distance.

The heated debate surrounding authentic musical performance peaked in 1980s. With seminal texts such as *Authenticity and Early Music*, the proceedings from a ground-breaking symposium of the same name, edited by Nicholas Kenyon, leading musicologists and performers in the 1980s rethought the authenticity enterprise. It took its place among other important studies including the philosopher, Peter Kivy's contribution *Authenticities; Philosophical Reflections on Musical Performance*.[48] The breadth, complexity and sophistication of this discourse, developed particularly in specialist journals such as *Early Music*, brought to the fore central questions about authenticity. This sustained process of self-reflection and self-examination, and the continued questioning of motives and agendas has been enormously constructive and has had a liberating effect on the playing itself.

Of these many commentators, Richard Taruskin has been perhaps the most polemical and controversial figure. Taruskin has deftly and engagingly shown that past generations of the early music revival brought to it their own modern sensibilities. He has argued that Early Music's text centricity, the emphasis on formal clarity and precision, the eschewal of profound sentiment and spontaneity, the striving to be impersonal, the fleet tempi, the small ensembles 'represented neither any determinable historical prototype nor any coherent revival of practices coeval with the repertories they addressed' but rather 'embodied a whole wish list of modern(ist) values'.[49] As a ferocious critic of musical modernism, Taruskin flayed the early music movement for what he saw as its 'brute modernist bias', and its 'civilising, sanitizing, miniaturizing project'.[50] He called for an end to the 'pretense' that what historically informed performers were doing was 'historically correct'. That they were not 'ransacking history in pursuit of truth', he claims, but were seeking permission. This deference to the authority of the printed score was part of a more general problem brought about the tyranny of *Werktreue*. He has argued persuasively that old music has been shaped to please our modern tastes and that this is not only inescapable but how it should be; it 'is the sound of now not then'.[51]

Taruskin was instrumental, although by no means alone, in the push beyond the naïve assumption that what historically informed performers did was 'real' in the sense of authentically reproducing a past event. Many now acknowledge that it is a real, and therefore authentic, attempt of their particular present's interpretation of the past. While discussing sixteenth-century madrigals and masses, Collingwood determined that 'Ancient art does not become modern art simply by being performed over again'. [52] Some musicians would argue that ancient art does indeed become modern when performed again. Kerman is only one of many to note that early music performance is 'a unique, difficult blend of old and new, a play of the contemporary creative sensibility upon the past'.[53] Musical performance, like reenactment, offers an immersive and affective experience that is potentially a rich avenue of discovery. In order to analyse and critically engage with the event and individual responses to it, the scholar and critic should maintain a sense of historical distance, in effect to hold on to the present. Collingwood himself exhorted the historian (not the musician) not to lose sight of historical distance as he continued his thoughts on sixteenth-century music:

> It is both interesting and delightful to sing madrigals and masses of the sixteenth century; but the historian is well aware, when he sings them and hears others sing them, that their place is in the sixteenth-century and not in the twentieth. He listens to them not simply as music – not simply as the expression of feeling in musical

> language – but as sixteenth-century music belonging to a bygone world whose mind and civilization he is trying to understand.[54]

Perhaps it was the seductive power of musical performance that caused him to cut all mention of music in his later writing on reenactment. He sensed the dangers inherent in embodied reenactment when he wrote that 'historical thinking is an activity ... which is a function of *self-consciousness*, a form of thought possible only to a mind which *knows itself to be thinking in that way*'.[55] The self-awareness that Collingwood calls for is a kind of disinterest on the part of the historian even while the reenactment is occurring and is akin to what Iain McCalman has called 'reflexive reenactment'.[56]

Despite the expressiveness of music, the question of affect – of emotion and expression – in historical performance has been fraught and complicated, and it has always unavoidably reflected the present. Dolmetsch and his contemporary Early Music pioneer, the Polish harpsichordist, Wanda Landowska, rejected what they saw as the bombastic, grandiose and simply too-loud music of the nineteenth century. They, like Collingwood, sought another sensibility, one simpler and clearer. However, their attitudes toward expression in ancient music, premised as they were on the power of intuition and spiritual empathy, were inescapably of their time. Both sought to capture the spirit' of the music. Dolmetsch wrote in the introduction to *The Interpretation of the Music of the Seventeenth and Eighteenth Centuries*:

> But the student should first try and prepare his mind by thoroughly understanding what the Old Masters *felt* about their own music, what impressions they wished to convey, and, generally, what was the *Spirit of their Art*, for on these points the ideas of modern musicians are by no means clear.[57]

For Dolmetsch and Landowska then, authenticity was achieved through a true representation the 'spirit of the composer.' Landowska famously said 'You play Bach *your* way, I'll play Bach *his* way'. She believed in her power to call forth 'the melodious echo' which united 'sympathetic hearts across the centuries'.[58] Despite Dolmetsch's efforts to affectively convey the spirit of the old music, his performances were nonetheless criticized for a lack of expression.

After the end of World War Two, the historical performance movement was increasingly dominated by the positivism of *Werktreue*. This approach, which incidentally was not restricted to early music movement but was part of a wider phenomenon, had a curious and paradoxical effect on early music performance. Some early music practitioners self-consciously refused to express emotion as if it would somehow taint the authenticity of the historical reenactment. They would simply let 'the music speak for itself'. [59] Perhaps believing themselves to be the most inauthentic ingredient of the performance, they removed themselves as far as possible from the event. They did not express, distrusting perhaps, as Dreyfus has claimed, 'their own intuitive impulses as a harmful residue of a Mainstream upbringing'.[60] Their approach was infused with reticence and restraint. It was a matter of taste; a corrective to the excesses of the late nineteenth-century. Erwin Bodky makes this explicit in a contribution to the Cambridge Society for Early Music in 1950 in which he characterized the movement as 'highly aristocratic, governed by restraint' in 'the display of emotion as well as the exhibition of technical virtuosity'.[61]

This reluctance to express emotion has been seen by Taruskin, Dreyfus and others as having its roots in early modernist aesthetic sensibility. Emotional restraint coloured the affective palette of early twentieth-century music such as French neoclassicism and Germany's *Die Neue Sachlichkeit*. Stravinsky and Ravel famously declared that they did not want their music to be 'interpreted' but merely played. This restraint was felt in historically informed performance. The subsequent generations assumed a more deliberately 'objective' approach which epitomized the modernist eschewal of the extravagant sentimentalism of the nineteenth century. It was a type of resolutely anti-sentimental reenactment and did not appeal to all. Dreyfus bemoaned the 'wretched renditions ground out by stony faced champions of authenticity'[62]

The recent affective turn experienced in the writing of history has also been felt in early music, which now allows for a more individualistic, expressive approach. The performer has to a greater extent stepped back in; the 'sewing machine Bach' of the 1950s has given way to a much freer, expressive approach. As early as 1992, John Rockwell compared the 'recent penchant of many of our livelier young early musicians towards greater flexibility, ornamentation and spirited play in their performances' with the '"objective", rigorous, briskly metronomic performances of the 1950s, 60s and 70s'. The recent generation showed, he observed, a greater interest in 'fancy, wit, rhetorical drama'.[63] Twelve years later, in 2004, Kenyon joined Rockwell in welcoming what he called 'the post-modern age of performance where all sorts of not-precisely-correct approaches can offer their own revelations'.[64]

Mark Salber Phillips has suggested that historians of recent generations have wanted to know more urgently how the past really felt.[65]

Questions of affect can extend into the phenomenological and sensory realm: we can ask how it was actually seen or heard. How did it sound? How did they listen? What emotions were aroused in the performer and the audience? These questions have been contemplated for decades in the field of historical performance. In 1950, Putnam Aldrich presciently pointed out that 'The performer of Baroque music ... finds that what he needs most in his struggle towards authenticity is the ability to equip his audience with Baroque ears'.[66] Aldrich's dilemma reminds us that even if the reconstruction is historically correct, our 'listening habits and expectations' are irreconcilably dissimilar to the time from which the music came.[67] We cannot play or hear music without the filters of all the music heard in the intervening years. Nor can we assume people from the past had similar emotional responses to or even heard sound as we do. The major scale has different affective properties for us than its equivalent did for Monteverdi and his contemporaries. Just as Umberto Eco has shown the ancient Romans did not see colours as we do, neither may we hear as they did.[68] Certainly we cannot assume the words used to describe their experiences match current meanings.

A musician who wishes to produce historically informed performances of past music has to have not only a deep understanding of past musical theories of affect, but also their relation to the sonic potential of the original instruments. The capabilities of the instrument determine musical expression. The two are bound up with each other and their relationship can illuminate the production of affect as sound. From the outset Dolmetsch demonstrated that past music 'could not be fully understood without reference to the sonorities of the instruments on which it was originally played and the performance practice of the period in which it was written'.[69] Although Dreyfus does not believe that Early Music preaches 'some empathetic leap into the past in an act of imaginative Verstehen', he does believe that a deep consideration of historical context, in addition to a study of performance manuals, will allow a more profound understanding of the temper of the age. [70] To limit the composers' intentions to the musical score is to strip them of their historicity; to wrest them from their world. It may be impossible to have an 'authentic' affective relationship with the past, however the act of trying to form such an emotional connection can reveal much about that particular instance both past and present.

In 2007, Vanessa Agnew alerted us to the pitfalls of collapsing temporalities in trying to bring the past nearer through reenactment.[71] Musicologist Carl Dahlhaus was aware of the dangers much earlier. He warned in 1983 that 'there is no escape in sight from the dilemma that to feel close to things past is to misconstrue them, while to understand them is to sense their remoteness'.[72] Historically informed performance has in many cases sought not to narrow the distance between past and present but rather to emphasize it. This is particularly noticeable in the historical performance incursion into what has long been considered canonic mainstream repertoire – the music of Mozart, Haydn, Beethoven, Schubert and Brahms. It is encroaching ever further forward. Familiar works are reproduced using sounds so unfamiliar to the modern ear so as to render the piece quite strange. We are forced to acknowledge that music so familiar to us was written centuries ago. Performing a work on original instruments and using historical performing practices, rather bringing it nearer, can have the opposite effect of making it seem further away. It is to render it once again strange and new. The Russian formalist Viktor Shklovsky's concept of *ostranenie* or defamiliarization is useful here. Although he applied it to the notion of the everyday in a literary context, it also can work within an historical framework.[73] Reenactment as process can do the same to more conventional modes of historical inquiry. But an historically informed musical performance is not only a process, research tool and mode of communication, but *itself* provides an important *source* for the historian.

Twentieth-century performances of early music have become important documents revealing important shifts along a continuum of time. The vast body of early music recordings spanning the last century provides the music historian with an invaluable resource of information about the tastes and sensibilities of that particular present and its understanding of that particular past. These archives of recorded sound allow mappings and interpretations of these different positions. Using recorded sound we can hear what audiences of the early twentieth century heard. Often it is surprising how unusual these early performances sound. Early music specialist, the renowned conductor Roger Norrington, has recently called for research into orchestral performing practices of this period, saying 'I don't see how we can continue to encourage musicology about 19th-century orchestral music when we have hardly begun to reconsider what it sounded like'.[74]

Ironically Collingwood's experience of Beethoven symphonies would have been through just such a nineteenth-century orchestral sound. He himself never would have 'heard an orchestra of the Beethoven period' that he deemed vital to obtaining a 'good, imaginative hearing' of a Beethoven score. The authentic restoration of that era did not begin during his lifetime. This raises the thorny issue of the value, or even possibility, of restoring lost or broken traditions. Nicholas Temperley observed that 'Adopting authenticity as the ultimate criterion would lead to the conclusion that medieval music, for example, cannot be performed at all, for lack of historical data'.[75] Haskell has also noted that the chances of authentically reproducing medieval music are slight, 'even if one has version of the score', for it was 'written in a culture of orality and improvisation' and so requires 'a reinvention more than a reconstruction'.[76] Consequently some have deemed it a useless enterprise. As an elderly composer, Rossini, when asked about the possible revival of *bel canto* singing replied emphatically, 'How can you imagine that it is possible to resuscitate what is dead? What genius ... could succeed in reconstituting that which has no tradition but oral tradition?'[77] The conductor Christopher Hogwood similarly rejected pre-seventeenth-century music because of unreliable sources – he could not be sure what he was doing was authentic. Too much depended on '*feeling*'.[78] Others subsequently have believed it worth pursuing. Taruskin exhorted performers to fill gaps with their imaginations, saying provocatively, that it was'fine to assemble the shards of a lost performance tradition, but how much better to reinvent it'.[79] Clearly creative exchange with the past in this instance is imperative.

Like historical reenactment, historical performance has been dismissed as nostalgic by its detractors. Theodor Adorno attacked the movement for its 'impotent nostalgia'. He and Nietzsche, also a detractor, believed the old music had to be adapted and reinterpreted for modern ears. Nietzsche believed that 'really historical performance would talk to ghosts'.[80] That it found such ardent admiration in the literary and artistic British avant-garde says something very interesting about that moment in history. Their strong urge to 'inhabit the past' was brought about by their need for refuge from the present.[81] Early music, despite its age, provided novel sounds and novel instruments but in a musical language that was familiar, unlike the music actually being produced at the time. For, as Dreyfus reminds us, Early Music arrived with atonality: 'the most profound crisis in European musical culture', one which was 'rejected by the middle-class public'. 'Early Music', he goes on to suggests, 'redressed the imbalance by repressing the nightmarish present and mounting a grand restoration of the glorious past'.[82] Despite this very real problem, many have noted that some of the most interesting and thoughtful contribution to musical performance have come from the historical performance world. In 1990, Leo Treitler pronounced historical performance as representing 'the most enlightened and enlightening thinking about the relation between music and history'.[83]

For Charles Rosen, it has enriched musical life. Even Taruskin grudgingly concedes it is the 'least moribund aspect of our musical life'.[84] Many modern orchestras and conductors have taken up ideas and techniques from historically informed performance practice.[85]

The contradiction between Collingwood's early and later writing on reenactment is telling. That the source of this contradiction is found in his discussion of music is also significant. His treatment of music in 1928 reveals that the physical experience, the reenactment of the past as lived experience, was important to him. It lends weight to a scholarly interest in performative, embodied reenactment, one that actualizes the idea. In his *Autobiography*, Collingwood speaks of himself as three entities. Significantly, his favourite was the third 'R. G. C.': the 'man of action', or rather, the one 'in which the difference between thinker and man of action disappeared'.[86] Similarly the early music movement is a marriage of thought and action; a field that has (to borrow Curtis Price's words), erased 'the boundary between scholarship and performance'.[87] As such it may provide a useful heuristic for academic historians interested in the possibilities of reenactment.

Collingwood theorized about historical reenactment during an age of revivals; an age suffused with nostalgia and preoccupied with reconstruction, reconstitution, restoration and reinvention. This rejection of modernity in favour of an idealized past was brought about by a sense of alienation from the present. Reenactment thrives on alienation. For Arnold Dolmetsch, the alienation was so profound and the desire to escape into the past so acute that he re-enacted the past every day through musical performance. This reactionary and retrogressive step was far from impotent. It was central in the development of a movement of such energy and influence that even its critics have acknowledged it as one of the most novel and innovative forms of contemporary musical performance.

Notes

1. R. G. Collingwood (1970 [1939]) *An Autobiography* (London: Oxford University Press), p. 3.
2. F. Neumann (1989) *New Essays on Performance Practice* (Ann Arbor; London: UMI Research Press), p. 3.
3. See L. Goehr (1992) *The Imaginary Museum of Musical Works: An Essay in the Philosophy of Music* (New York: Oxford University Press).
4. L. Dreyfus (1983) 'Early Music Defended against Its Devotees: A Theory of Historical Performance in the Twentieth Century', *Musical Quarterly*, 69:3, p. 298.

5. W. M. Johnston (1967) *The Formative Years of R. G. Collingwood* (The Hague: M. Nijhoff), p. 29.
6. Collingwood, *An Autbiography*, p. 13.
7. Ibid., p. 7.
8. Johnston, *The Formative Years*, p. 14.
9. R. G. Collingwood (1994) 'Outlines of a Philosophy of History' in *The Idea of History with Lectures 1926–1928*, ed. Jan Van Der Dussen (Oxford: Oxford University Press), pp. 438–95.
10. Collingwood, *An Autobiography*, pp. 109–14.
11. Collingwood, 'Outlines of a Philosophy of History', p. 441.
12. In his study of Collingwood's idea of re-enactment, William Dray notes that a 'peculiarity of his presentation of it at that point [in 1928] is his reporting that the idea first came to him while asking himself how one understands the present performance of a piece of music composed at some earlier time'. See W. Dray (1995) *History as Re-enactment: R. G. Collingwood's Idea of History* (Oxford: Clarendon Press), pp. 33, 138–9.
13. D. H. Johnson (1994) 'W. G. Collingwood and the Beginnings of the Idea of History', in David Boucher (ed.), *The Life and Thought of R. G. Collingwood, Collingwood Studies* 1 (Swansea: R. G. Collingwood Society), p. 1.
14. E. Pound (1954) 'Arnold Dolmetsch', in T. S. Eliot (ed.) *Literary Essays of Ezra Pound* (London: Faber), p. 434.
15. F. Howe (1949) 'Dolmetsch, (Eugene) Arnold', *Dictionary of National Biography Archive*, available at http://www.oxforddnb.com/view/article/32853.
16. Ezra Pound recommended this book for inclusion in the Little Review Bookshop. See T. L. Scott, Melvin J. Friedman and J. R. Bryer (eds) (1988) *Pound/The Little Review: The Letters of Ezra Pound to Margaret Anderson: The Little Review Correspondence* (New York: New Directions), p. 113.
17. H. C. G. Matthew and J. Craig-McFeely, 'Dolmetsch, (Eugène) Arnold (1858–1940)', *Oxford Dictionary of National Biography*, available at http://www.oxforddnb.com/view/article/32853.
18. M. Campbell (1975) *Dolmetsch: The Man and His Work* (London: Hamish Hamilton), p. 68.
19. D. H. Laurence (ed.) (1981) *Shaw's Music: The Complete Musical Criticism in Three Volumes*, vol. 3 (London: Max Reinhardt, The Bodley Head), p. 162.
20. Laurence, *Shaw's Music*, vol. 3, p. 262.
21. Ibid., p. 178.
22. Ibid., p. 332.
23. Haskell, *The Early Music Revival*, p. 29.
24. Laurence, *Shaw's Music*, vol. 3, p. 332.
25. F. MacCarthy (1994) *William Morris: A Life for Our Time* (London: Faber), pp. 668–9.
26. Roger Fry's designs featured on some of Dolmetsch's instruments.
27. H. Haskell (1988) *The Early Music Revival: A History* (London: Thames and Hudson), p. 29. He also had a reputation for being difficult and impatient.
28. Ibid., p. 30.
29. Scott et al., *Pound/The Little Review*, p. 188.
30. Ibid., p. 186.
31. Pound, 'Arnold Dolmetsch', p. 435.
32. E. Pound (1985) *I Cantos* (Milan: Arnoldo Mondadori Editore), p. 1018.

33. Cited in Haskell, *The Early Music Revival*, p. 30.
34. J. Joyce, (1936) *Ulysses* (London: J. Lane), p. 1082.
35. Campbell, *Dolmetsch*, p. 70.
36. G. Moore (1898) *Evelyn Innes* (London: T. Fisher Unwin), available at http://www2.cddc.vt.edu/gutenberg/1/3/2/0/13201/13201-h/13201-h.htm#CHAPTER_ONE.
37. *Westminster Gazette* (15 November 1893) cited in Campbell, *Dolmetsch*, p. 57.
38. Campbell, *Dolmetsch*, pp.65–6 (emphasis added).
39. Haskell, *The Early Music Revival*, p. 31.
40. Cited in B. Blood 'The Dolmetsch Story', available at http://www.dolmetsch.com/Dolworks.htm.
41. J. Bloomfield (5 July 2007) 'William Poel's Shakespeare: The First Historically Authentic Productions', available at http://shakespearean-performances.suite101.com/article.cfm/william_poels_shakespeare.
42. Neumann, *New Essays on Performance Practice*, p. 10.
43. Haskell, *The Early Music Revival*, p. 43.
44. Wimsatt and Beardsley's concept of the intentional fallacy has been used by those problematizing this approach. See for example, L. Treitler (2003) *With Voice and Pen: Coming to Know Medieval Song and How It was Made* (Oxford: Oxford University Press).
45. Haskell, *The Early Music Revival*, p. 175.
46. J. Kerman (Winter 1992) 'The Early Music Debate: Ancients, Moderns, Postmoderns', *Journal of Musicology*, 10:1, p. 113.
47. N. Kenyon (August 2004) 'The Historical Imagination', *Early Music*, 32:3, p. 459.
48. See N. Kenyon (ed.) (1988) *Authenticity and Early Music* (Oxford: Oxford University Press); P. Kivy (1995) *Authenticities: Philosophical Reflections on Musical Performance* (Ithaca: Cornell University Press).
49. R. Taruskin (1995) 'Last Thoughts First' in *Text and Act: Essays on Music and Performance* (New York: Oxford University Press), p. 5.
50. Ibid., p. 6. His ferocity has coloured his titles. See for example, R. Taruskin (1984) 'The Authenticity Movement can become a Positivist Purgatory, Literalistic and Dehumanizing', *Early Music*, 12:1, pp. 3–12.
51. R. Taruskin, 'The Modern Sound of Early Music', in *Text and Act*, p. 166.
52. Collingwood, 'Outlines of a Philosophy of History' p. 443.
53. J. Kerman (1985) *Musicology* (London: Fontana Press), p. 200.
54. Ibid.
55. Collingwood, *Idea of History*, p. 289 (emphases added).
56. I. McCalman 'Past Not Just the Present Dressed in Funny Clothes', *Sydney Morning Herald* (17 September 2007), available at http://www.smh.com.au/news/opinion/past-not-just-the-present-dressed-in-funny-clothes/2007/09/16/1189881340166.html.
57. A. Dolmetsch (1969) *The Interpretation of the Music of the Seventeenth and Eighteenth Centuries*, 3rd edn (Seattle: University of Washington Press), p. vii.
58. W. Landowska (1924) *Music of the Past* (New York: Alfred A. Knopf), p. 175.
59. T. Livingston (1999) 'Music Revivals: Towards a General Theory', *Ethnomusicology*, 43:1, p. 76.

60. Dreyfus, 'Early Music Defended against its Devotees', p. 319.
61. Haskell, *The Early Music Revival*, p. 178.
62. Dreyfus, 'Early Music Defended against its Devotees', p. 304.
63. J. Rockwell (Winter 1992) 'The Early Music Debate: Ancients, Moderns, Postmoderns', *Journal of Musicology*, 10:1, p. 123.
64. Kenyon embraced these 'postmodern' performance for offering 'new freedoms for the way we absorb, internalize and re-create historical information' and identified them as 'a fruitful subject for the next decade of research. In 2007, Nancy November identified 'a newer attitude' in historical performance, one that 'recognizes that the task of "cleaning away" performance traditions is impossible and that such an attempt is even undesirable'. In the same year, Bruce Haynes devoted a monograph to historically informed performance with a similar thesis at its core. See Kenyon, 'The Historical Imagination', p. 459–60; N. November (2007) 'The Emperor's New Clothes?: "Period" Beethoven and Performance Traditions', *Early Music*, 35:3, p. 488; B. Haynes (2007) *The End of Early Music: A Period Performer's History of Music for the Twenty-First Century* (Oxford: Oxford University Press).
65. M. Salber Philips (29 May 2007) 'On the Advantages and Disadvantages of Sentimental History for Life', seminar paper given at the Humanities Research Centre, Australian National University.
66. D. Fabian (2001) 'The Meaning of Authenticity and the Early Music Movement: A Historical Review', *International Review of the Aesthetics and Sociology of Music*, 32:2, p. 159. In 1970, Karl Gustav Fellerer asked: 'With which type of sound can contemporary man experience the historical art work in the way the person of that experienced it?' 'How', he continued 'can the historical art work be made to react on contemporaries in the same manner as it did then?' Cited in Fabian, 'The Meaning of Authenticity and the Early Music Movement', p. 165. During the eighties and nineties, Israeli musicologist, Shai Burstyn pursued this idea of authentic listening and the 'period ear' in several articles. See for example S. Burstyn (1997) 'In Quest of the Period Ear', *Early Music*, 25:4, pp. 692–7, 699–701; S. Burstyn (1986) 'Authentic Listening?', *Orbis musicae*, ix, pp. 141–9; S. Burstyn (1998)'Music as Heard', *Early Music*, 26:3, pp.515, 517–18. (The latter is a report on the conference 'Music as Heard: Listeners and Listening in Late Medieval and Early Modern Europe (1300–1600)' held at Princeton University in September 1997); S. Burstyn (1998) 'Pre-1600 Music Listening: A Methodological Approach, *Musical Quarterly* (Special Issue: 'Music as Heard'), 82:3/4, pp. 455–65. See also L. Botstein (1998) 'Toward a History of Listening', *Musical Quarterly* (Special Issue: 'Music as Heard'), 82:3/4, pp. 427–31.
67. Fabian, 'The Meaning of Authenticity and the Early Music Movement', pp. 165–6.
68. See U. Eco (2004) 'Lingue Perfette e Colori Imperfetti', in *Dire Quasi La Stessa Cosa* (Milan: Bompiani), pp. 345–53. I am grateful to Geoff Murrell for this reference.
69. Haskell, *The Early Music Revival*, p. 43. Michael Morrow suggests widening this approach, believing that 'by reading the classics, looking at paintings and sculpture and attending theatre ... we can share – though in a limited manner – the feelings, emotions and ideas of men of other ages and other

civilizations'. 'But', he hastens to remind us, 'we must never forget that in any age the artist is addressing himself to his contemporaries, and his language is composed of a system of familiar conventions – musical, visual or literary. If we don't or can't learn these languages, the conventions will be as meaningless to us as the hand gestures of an Indian dancer are to the average western audience'. See M. Morrow (1978) 'Musical Performance and Authenticity', *Early Music*, 6:2, p.233.

70. Dreyfus, 'Early Music Defended against Its Devotees', p. 299.
71. V. Agnew (2007) 'History's Affective Turn: Historical Re-enactment and Its Work in the Present', *Rethinking History*, 11:3, p. 299.
72. C. Dahlhaus (1983) *Foundations of Music History* (Cambridge: Cambridge University Press), p. 63.
73. For an elucidation of this idea of estrangement, which has much in common with the Brecht's concept of *Verfremdung*, see V. Shklovsky (1990) *Theory of Prose*, trans. Benjamin Sher (Elmwood Park, IL: Dalkey Archive Press).
74. R. Norrington (February 2004) 'The Sound Orchestras Make', *Early Music*, 32:1, p. 2. There have been sustained studies in this area. See for example, R. Philip (1992) *Early Recording and Musical Style, 1900–1950* (Cambridge: Cambridge University Press); E. Ornoy (2006) 'Between Theory and Practice: Comparative Study of Early Music Performances', *Early Music*, 34:2, pp. 233–47. A new archive of early music recordings was established in 1990 at the Early Music Institute, School of Music, Indiana University, The aim was 'to preserve in sound the development in the 20th century of historical performance practices (including the uses of original instruments) for music of the Medieval, Renaissance, Baroque, and later periods'. See (1990) 'New Archive of Early Music recordings', *Early Music*, 18:3, p. 506.
75. N. Temperley (February 1984) 'The Movement Puts a Stronger Premium on Novelty than on Accuracy, and Fosters Misrepresentation', *Early Music*, 12:1, p. 18.
76. Haskell, *The Early Music Revival*, p. 176–7. See also D. Leech-Wilkinson (2002) *The Modern Invention of Medieval Music: Scholarship, Ideology, Performance* (New York: Cambridge University Press).
77. R. Wistreich (2007) 'Lost Voices', *Early Music*, 35:3, p. 457.
78. Fabian, 'The Meaning of Authenticity and the Early Music Movement', p. 155 (emphasis added).
79. Dreyfus also prescribed 'a large amount of musicianship and invention' as a means of bridging the gap between historical evidence and a sounding performance. See Dreyfus, 'Early Music Defended against Its Devotees', p. 13.
80. Haskell, *The Early Music Revival*, p. 177. Later in the 1980s, Taruskin still warned of the dangers of 'time-travel nostalgia'. See R. Taruskin (1982) 'On Letting the Music Speak for Itself: Some Reflections on Musicology and Performance', *Journal of Musicology*, 1:3, p. 342.
81. In her study of music revivalism, Tamara Livingston points out that the Early Music Revival attracted a disaffected middle class who 'felt alienated by urban life' and the cultural mainstream, and were drawn to oppositional, reactionary ideology. See Livingston, 'Music Revivals', p. 74.
82. Dreyfus, 'Early Music Defended against Its Devotees', p. 305. Robert Morgan, a specialist in twentieth-century music, understands the rejection of new music for new ways of looking at the past as expressing a general

anxiety about the state of contemporary art music. Historical performance movement, he believes, has in effect usurped the place of new music. See R. Morgan (1988) 'Tradition, Anxiety, and the Current Musical Scene' in Kenyon (ed.), *Authenticity and Early Music*, pp. 57–82. Joshua Kosman also believes that historical performance has, 'substituted archaeology for new creation'. See J. Kosman (Winter 1992), 'The Early Music Debate: Ancients, Moderns, Postmoderns', *Journal of Musicology*, 10:1, p. 119.

83. L. Treitler (1990) 'History and Music', *New Literary History*, 21:2, p. 318.
84. B. D. Sherman (2003) *Inside Early Music: Conversations with Performers* (Oxford: Oxford University Press), p. 7.
85. Kenyon notes that historical performance has 'infiltrated mainstream performance' and broadened its expressive and timbral palette. See Kenyon, 'The Historical Imagination', p. 450. Stanley Ritchie makes a similar observation that mainstream performers have been, 'influenced by the revelation of earlier esthetic principles, and although it might not always be practical for them to play on original instruments, they are often open to revising certain interpretative ideas, such as tempo choice, phrasing style and use of the long-forgotten art of embellishment'. See S. Ritchie (Autumn 1984) 'Authentic Reconstruction of Musical Performance: History and Influence', *The Drama Review*, 28:3, p. 73.
86. Collingwood, *An Autobiography*, p. 151.
87. C. Price (November 1997) 'Early Music: Listening Practice and Living Museums', *Early Music*, 25:4, p. 561.

10

'From Wigwam to White Lights': Popular Culture, Politics, and the Performance of Native North American Identity in the Era of Assimilationism

Ruth B. Phillips and Trudy Nicks

> For everyone involved in reenactments, it is a transformative experience, for it allows us to momentarily step into a real or imagined past through a political or cultural lens, never the historiographic route. Thus, the double entendre of reenactment: when elders are prone to say, 'It's hard to be an Indian', we now know it plays both ways for Natives and non-Natives.
>
> —Gerald McMaster, 2007[1]

The major collection of historical objects owned by the Kanien'kehaka Onkwawén:na Raotitiohkwa, the Mohawk cultural centre at Kahnawake, Quebec outside of Montreal, is an assemblage of clothing and memorabilia that testifies to the strategic nature of Aboriginal reenactments of Native North American identity during one of the darkest periods in indigenous colonial history.[2] The collection is the legacy of Esther Deer, an acclaimed Hodenosaunee (Iroquois)[3] professional entertainer who performed in Europe, Africa, and North America under the name of Princess White Deer between the late 1890s and the mid-1930s. It includes publicity photographs, scrapbooks of press clippings, theatrical programs, official documents, correspondence, and souvenirs of Deer's international travels. The stage costumes in the collection include finely beaded hide garments typical of 'authentic' nineteenth-century Plains Indian clothing, fantastically coloured feather bonnets (which Deer wore, as her publicity photos show, with the tiny spangled halter tops and briefs also in the collection), and fashionable dresses in the flapper style of the 1920s.

As the material assemblage of a twentieth-century Mohawk woman, Esther Deer's clothing, photographs, and mementos contrast sharply with Hodenosaunee collections in non-Native anthropology and art museums and archives. If her collection had been offered to one of these institutions, the beaded hide garments would almost certainly have been transferred to the Plains collection, the souvenirs to decorative arts and African ethnography, while the theatrical costumes would probably have been rejected outright or left in the storage cabinets under a penumbra of dubious authenticity.[4] At the Kanien'kehaka Onkwawén: na Raotitiohkwa, however, the heterogeneity of the Princess White Deer collection is not problematic. Community members remember with pride – and often with a dry, self-reflexive humour – the numerous forebears who have left over the years to work as popular entertainers and re-enactors of stereotypical Plains warriors and Indian princesses.

The artifacts, depictions, ephemera in the Princess White Deer collection are the material and visual deposit left by performances that we can no longer witness and for which we have no scripts or scores and only the most fragmentary descriptions. In this essay we read the Princess White Deer collection as a visual and material assemblage in order to supply evidence of indigenous agency which cannot be retrieved from the standard textual sources. We will argue that it testifies to Esther Deer's use of reenactment to destabilize potentially paralyzing stereotypes of Indianness current during the early twentieth century. This period was, arguably, a time when Indian identities came under greater threat than ever before in post-contact history. During these years the doctrine of cultural evolutionism and the belief that distinctive Native cultures could not survive the forces of modernization and would soon disappear were almost universally accepted by non-Indians.[5] In both the US and Canada, governments aggressively enforced laws and policies designed to engineer the erasure of Native languages and cultures and the assimilation of indigenous peoples into settler society. Amidst such pressures, as we will also argue, Esther Deer adapted her reenactments and made calculated choices of self-presentation through dress and performance style in relation to the different sites in which they were staged, from Wild West shows, to patriotic war rallies, vaudeville, and campaigns for Indian rights. Like other indigenous entertainers of the period, she exploited the capacity of reenactment, as a performative genre, to accomplish complex processes of cross-appropriation to counter the assumption that authentic American Indian identities were exclusively tied to an unrecoverable past and to draw attention to the modern lives Aboriginal people were actually living.

Indian stereotypes and scholarly critiques

Non-Native and Native post-colonial scholars have tended to interpret Indian participation in popular performance differently. In several foundational post-colonial analyses non-Native writers represented Native reenactments as essentially passive. Literary scholar Terry Goldie, for example, concluded in *Fear and Temptation* (1989), that 'the indigene is a semiotic pawn on a chess board under the control of the white signmaker',[6] while historian, Daniel Francis, wrote in *The Imaginary Indian* (1992), that 'the Performing Indian was playing a role in somebody else's story'.[7] Such remarks imply that Indian stereotypes constituted a closed and endlessly self-reproducing system which reenactments served only to re-inscribe. Selling Indianness, according to this view, was selling out.

A number of Aboriginal scholars have sought to complicate such readings. In *Fluffs and Feathers*, a ground-breaking exhibition about stereotypes of Indianness in 1988, the late Mohawk historian Deborah Doxtator addressed stereotypes of Indianness in advertising, film, and other aspects of popular visual culture.[8] Her argument, that reflexivity, irony, and humour have long characterized Aboriginal reception of and participation in stereotypical representations, has since been supported by other Aboriginal commentators. Recalling Doxtator's exhibition, Anishinaabe writer Lenore Keeshig-Tobias both concurred with, and provided further evidence for, Doxtator's argument. 'As the curator for this incredibly funny and satirical exhibit of mostly non-Aboriginal symbols of Indianness', she recalled, 'Deborah had pulled together posters, city crests, biscuit tins, food labels, and all kinds of kitsch and bric-a-brac; horribly funny Indian stuff. It was fun'.[9]

In a more recent curatorial project assessing the history of Indian reenactments in film and contemporary visual art, Aboriginal curator and artist Gerald McMaster has further developed this revisionist position. 'To some extent', he argues, 'these plays, like the shows and films, affirmed an aboriginal identity, giving its performers a brief, albeit emancipatory, moment, one that was however based on anachronism (past)'.[10] McMaster also considers the problem of non-Native reception, asking:

> Why would this fantasy role-playing with its appropriation of an imagined past appeal to viewers? In a way there was sympathy for these dislocated tribes. Though highly romanticized, the narratives were accessible to the audience's cultural frame of reference.

> The pageant plays were an entry point from which to understand aboriginal history via the tragic but noble terms from which they understood both Hiawatha and aboriginal peoples.[11]

Keeshig-Tobias's notion of the 'horribly funny' and McMaster's notion of the 'emancipatory moment' are consistent with Homi Bhabha's theorization of colonial mimicry as a means of exposing the partial representation of the stereotype, of producing ambivalence about its truth, and of intervening in colonial discourses about identity. 'In this comic turn from the high ideals of the colonial imagination to its low mimetic literary effects', he writes, 'mimicry emerges as one of the most elusive and effective strategies of colonial power and knowledge'.[12]

Rare but precious evidence of the subjectivity of early twentieth-century performers is provided by diaries kept by one of Esther Deer's contemporaries, the Penobscot dancer Molly Spotted Elk.[13] The thoughts and feelings she recorded as she performed in vaudeville, cabaret, and speakeasies were often couched in irony. On one occasion, for example, she wrote of some non-Native friends: 'It was sweet of the boys to come down to watch the Injun jump and do a few meaningless steps',[14] while in another diary entry, describing her act at the Frivolity Night Club, she wrote, 'Did my Indian number in the first show and the black bottom in the second ... Enjoyed it immensely'.[15] But the scanty and inauthentic clothing she had to wear frequently made her uncomfortable. 'My costume made me embarrassed. Looked like a loin cloth affair of satin and beads instead of leather and fringe. Not natural for my Indian dance ... I am just an injun in the flesh parade. Feel terrible about being bare and walking around, but I must work'.[16] Such passages recall Lakota literary critic Philip Deloria's observation that, 'using antimodern primitivism to defend native cultures against the negative stereotypes left over from colonial conquest, the interaction between Indians and non-Indians over the question of Indian Others proved extraordinarily complex, with material and ideal exchanges almost too subtle to grasp'.[17]

While Princess White Deer has left us no comparable first hand accounts, her theatrical costumes, props, publicity photos, and ephemera form a kind of visual analogue to Molly Spotted Elk's diary entries. As we will see, they suggest that her own choices were equally self conscious and strategic. Both performers took advantage of the co-temporality that characterizes performance genres, binding actor and audience through its shared experiential dimension and endowing with a unique immediacy the humour, irony, and mimicry implicit in reenactments of Indian stereotypes. In contrast with the more cerebral resonance of literary texts and graphic arts, performance thus produces a more visceral impact on both audience and performer. In looking at Esther Deer's biography and her work as entertainer and re-enactor, we will suggest how this interactive potential worked itself out in the fine grain of an individual life and career.

The famous Deer family: Champion trick riders of the world

Esther Deer was born into a family of professional entertainers, one of many Hodenosaunee from eastern reserves who, motivated by economic need, were active in the entertainment industry during the second half of the nineteenth- and early twentieth centuries.[18] Her father, James Deer, was born in 1866. He and his brother John were among the group of Iroquois boatmen who were recruited to join General Wolseley's British Army Expedition to rescue General Gordon at Khartoum in 1884–5 because of their expertise in piloting boats through the rapids of the St Lawrence River. By 1889 the Deer Brothers had founded a trick-riding troupe and were based in New York City, where James married a non-Native actress named Georgette Osborne. 'The Famous Deer Brothers, Champion Indian Trick Riders of the World' performed on their own and with larger Wild West shows.[19] Travelling in Europe with these larger spectacles during the pre-World War I years, the brothers also participated in reenactments of the Khartoum campaign.[20] Through her marriage, Georgette acquired legal status as an Indian and performed a wide range of daredevil acts as a member of the Deer Brothers' troupe. One programme describes her, for example, as 'jumping from one horse to another at full speed in the chase for a bride and many other difficult feats'.[21]

Esther Deer was born at Akwesasne in 1891 or 1892 and by the age of nine, she, too, was performing in the family troupe. She was a success at the Pan American Exposition in Buffalo, New York in 1904 as part of the Indian Congress, described in the official catalogue and guidebook as a gathering of '700 Indians representing 42 distinct tribes, and living in their aboriginal and primitive state. Fifteen famous War Chiefs, Indian Rough Riders, Indian Dances, Sports and Pastimes ... 500 Indians in a great realistic sham battle. No such lesson in history can be gleaned from books as is here shown by "Native Americans" – long-haired painted savages in all their barbaric splendour'.[22] Esther was also listed as a member of the troupe when it joined Texas Jack's Wild West Show on a tour of South Africa later that year. On the return voyage to London, 12-year old Esther entertained the shipboard passengers by singing Hiawatha, 'in such a captivating manner that she receives a double encore nightly'.[23] Between 1904 and 1910 the Deer Brothers performed primarily in Great Britain and on the Continent, and Esther probably acted in the historical dramatizations of 'Daniel Boone', and 'Deer Family, Indians of the Past' that were listed in the programmes of those years. The latter consisted of a series of 'Tableau in Two Scenes' including: (1) The Indian Camp Fire, (2) Princess White Deer in her Famous Songs and Indian War Dance, (3) The Settler's Cabin, (4) White Rose and Princess Deer in their Specialty, (5) The Indians Burning Settler's Cabin, (6) Sensational Knife Duel, and (7) Settlers to the Rescue.[24]

In 1913 and 1914, Esther and her mother were in Russia, performing in St Petersburg and Moscow. At a performance before the royal family, Esther received a gold medal 'for her clever dancing'.[25] In 1913 she met Count Alexis Krasicki, an officer in the Russian army, and married him two years later in England just after the outbreak of World War I. 'Princess' Esther thus became a countess, acquiring in reality the social status that had been coded in the stereotype of the noble savage which had first emerged soon after European contact with the indigenous peoples of the Americas and which had continued to gain currency throughout the following centuries. (Interestingly although she was referred to as a countess in newspaper articles for a brief period after her wedding, the stereotype of Indian 'Princess' proved more enduring in later years.) Esther Deer and her husband had only a few months together before Count Krasicki was killed in battle. Esther returned to North America soon after and never remarried.

Vaudeville, racial stereotypes, and 'Genuine Americanism'

By 1917 Esther Deer was back in New York City, working in the Keith-Albee circuit, which staged its vaudeville shows in the biggest and most elegant theatres of the day. The shift from Wild West shows to vaudeville entailed a shift of audience and milieu. While the Wild West shows were circus-like spectacles presenting Indians as 'long-haired painted savages in all their barbaric splendor',[26] to a broad cross section of the public, vaudeville was a more feminized and theatrical form of performance associated with refinement and middle-class gentility.[27] Vaudeville thus offered a milieu in which the persona of the Indian princess could serve as a primary point of reference for Indian identity.

As regularly noted in newspaper reviews of the period, Princess White Deer was also patriotically helping with the war effort. She entertained troops in the Liberty Theaters set up in a dozen different army camps, as well as giving Red Cross benefit performances for the war effort in New York. In this context the strong link between indigenous and American identities had a clear utility to the patriotic fervour required by the war effort. One newspaper quoted Esther as saying: 'I had to get back to my native land to be just an American girl and do my part in the war'.[28] In the same vein, a journalist in Savannah, Georgia described her as 'White Deer, a bright example of genuine Americanism', following a 1918 performance.[29]

Newspaper clippings dated to about 1917 describe these wartime spectacles in some detail. An article entitled 'Original Americans Head the Vaudeville Bill at the Grand' reported, for example, that

> Princess White Deer and Company, a trio of Indians, portrayed pastimes and customs of the original American in an unusual act ... It is elaborately presented. Two braves resplendent in war bonnets and beeded [*sic*] buckskin are watching the setting sun when the curtain rises. A camp fire in front of a teppee [*sic*] sends a soft glow over the scene harmonizing beautifully with rays of the dying sun.
>
> From the depths of the small teppee, comes a native song by a sweet voiced squaw. For several seconds the silvery tones made more beautiful by the steady pompom of the nearest approach Indians had to a snare drum continue, then an apparition in brown trips gracefully from the teppee and the audience witnesses one of the most graceful and weird exhibitions of dancing seen here in a long time.
>
> The call of the original American of today to duty to suppress German autocracy was recited by a brave with all of the solemnity of his people. He emphasizes the refrain that it is a privilege to die for one's country. The older of the braves chanted a weird song and there was more dancing by the princess, who performed with equal ease the blood curdling war dance and the clog of plantation darkies. The act is unusual, far above the average and worth going far to see.[30]

Through her work in the Liberty Theaters, Esther Deer came to the notice of Florenz Ziegfeld, whose famous Follies were a more elaborate theatrical form that had grown out of vaudeville. Ziegfield hired her to dance and act in pantomimes, and the elevation of her status that accompanied this success is suggested by a 1918 notice of a forthcoming show:

> The bill at the Bijou ... contains acts which are all dandies. Chief among them will be Princess White Deer and Co. genuine Indians.
> Princess White Deer of the Mohawks is an example of the social elevation possible for the redskin. She sings better than most girls sing and has mastered all of the modern dances. These and the dances of her tribe and its native music constitute the offering she is presenting in vaudeville.[31]

The reviews of this period also document the fact that White Deer, like virtually every female performer, participated in the currency of racial stereotyping that had been deeply entrenched in vaudeville since its inception. In 1921 a reviewer of Ziegfeld's 'Nine-O'Clock Frolic' reported that 'Princess White Deer exhibited a Dance to the Great Spirit that is bound to be imitated. She also showed that an Indian can copy from the Black race as well as the White by executing a difficult Buck and Wing Dance'.[32] Reviewers, however, consistently singled out the Indian acts for their special interest and quality (Figure 10.1). White Deer's act in 'Hitchy-Koo' was described as 'a ballet of real Indian girls who will dance like wild women from the Iroquois reservation'.[33] Another reviewer noted that this 'particularly clever Indian dance number [was] followed by a jazz selection as an encore'.[34] This is probably the same as the act in which, 'Capt. John Smith was carefully dusted and "Hitchy Kooed" and brought forth from the musty pages of history to a new jazz version of his ever-popular romance with the fair Indian princess, Pocahontas'.[35] The dizzying changes of persona within a single show seem to have been typical of White Deer's vaudeville career (Figure 10.2). For example, in her 1920 'Tip Top' show, Princess White Deer was seen in 'full Indian war dress to do a scalp dance, and later she wears modern evening dress and does a snappy buck and wing dance'.[36]

The 1920s: 'Jazzing with All'

In 1925 White Deer seems to have created a production of her own, entitled 'From Wigwam to White Lights: An Indian View', which was performed at several different New York theatres. This production is perhaps the most closely tied to conventions of historical reenactment. It was advertised as 'An American Indian Revue in Three Periods, 1776,

Figure 10.1 Princess White Deer in feather bonnet

1863, and 1918', and a review in *New York Variety* for 21 January 1925 describes it as 'an ambitious attempt for a novel dancing turn':

> The Indian ancestry of White Deer and the presentation are the outstanding features. The act is prolonged by a brave who eloquates ... about the 'daughter of the Mohawks'. The act is based upon dances of the 17th, 18th, and 19th centuries which serves as a skeleton for double dances of the Princess and a male dancing partner. Following the chief's address the act goes to full stage. A cyclorama parts, disclosing White Deer in native costume. She solos a symbolic dance, interpreting it gracefully. Cards at the side announce the next dance as of the 18th century. The Hip's chorus of eighteen girls follow in ante-bellum costume for a well executed minuet. White Deer and her partner, both in cowboy and girl costume, double a one-step and semi-jazz dance and fox trot. The 19th century brings on the chorus

Figure 10.2 Princess White Deer in oriental pose

> in Union Army costumes for a Virginia reel, also well done. The double of White Deer and her partner following has the pair in evening clothes for a novelty waltz. The 20th century dance introduces the chorus in West Point uniforms and dresses for a corking, kicking and dance drill. The Princess in the finish she is jazzing with all, working in an ensemble finale.[37]

'From Wigwam to White Lights' was so successful that it was booked as the headline show for the entire Keith-Albee circuit for a six-month run in 1925. By promoting the show as demonstrating 'the influence of the native American on art, music, dancing and the theater', its organizers made reference to the debt of modern Western art forms, including dance, to the arts of 'primitive peoples' which was widely acknowledged by the mid-1920s.[38]

Star status came to vaudeville performers at the cost of tremendous physical stress, and in 1925, in the midst of her success, Esther Deer's health broke down.[39] A trip to Florida to recuperate led to a new phase of her career. Partnered by Peppy De Albrew, also known as the Sheik of Brazil or the Argentine Sheik, she began to perform as a professional dancer at major hotels and private society parties. White Deer and Peppy performed the Charleston, the tango, and the waltz, as well as a new dance known as the Deer Stalk. This was described as 'an invention of the Princess and a ballroom dance descriptive of the spirit of the field and wood'.[40] Even in this new activity, Esther Deer continued to bridge Indian and non-Indian worlds. She was photographed in March 1925, within a few days of her arrival in Florida, in a Seminole Indian village teaching women how to put on makeup and dance the Charleston. In 1927 Esther was back in New York, dancing and acting in Ziegfeld's production of 'Lucky' at the New Amsterdam Theater, which was billed as 'her first real part with lines'.[41] A reviewer of this show referred to her as 'a snakey and fascinating siren who alone could answer the trite query as to why boys leave home'.[42] By 1928 she was on her way to Paris with a letter of introduction from the New Amsterdam Theater. She specialized there in the performance of Indian dances and, according to a newspaper article, was 'finishing a history of the Mohawk people, on which she has been working for two years'.[43]

Throughout these years, Esther Deer had continued to use performance as a way to bridge both Native and non-Native worlds by invoking the trope of the Native American as American patriot. In 1925 she was invited to perform a version of 'From Wigwam to White Lights' at the Testimonial Performance organized by the American Legion to honour General Pershing. As the invitation letter stated, her contribution was regarded as particularly appropriate because she was a 'REAL American'.[44] And in the same year the American Indian Defense Association, an organization which had been formed 'to secure for all Indians the right to land, to legal protection, health protection, modern education and liberty of conscience', invited her to appear for a second year at their Aztec Ball at the Ambassador Hotel.[45] When she performed a programme of American Indian songs on Boston's WGBS radio station in 1927, she was again promoted as 'a real "first American", being the granddaughter of the famous Mohawk chief, Running Deer'.[46]

Reenactments of the triumphalist settler historical narrative of American and Canadian history had been a staple of local and national commemorative celebrations since the nineteenth century, and they remained popular in the context of the new emphasis on greater cultural autonomy from Europe that characterized both Canada and the US during the post-World War I years. In 1933, for example, White Deer and her father appeared in a pageant held to commemorate the two hundred and fiftieth anniversary of purchase of the land now occupied by White Plains, New York. Her father impersonated Quarropas, the chief who sold the land in 1683 to the settlers of the town of Rye. Both wore Plains Indian dress; the beaded vest worn by James Deer and Princess White Deer's dress are both in the Kanien'kehaka Onkwawén: na Raotitiohkwa collection.[47]

In 1937, Esther wrote and recited a poem for the opening of a private resort community at Lake Mohawk, New Jersey which articulated the standard narrative of the peaceful surrender of Indian lands to a 'new race' that was a standard message of historical pageants. The poem she wrote for the occasion, 'To Lake Mohawk', read:

> And my people in the cloudland,
> looked below at sound of axes,
> Saw men building many houses,
> And new trails upon the hillside,
> Heard the whispers of the Woodland,
> Felt the joy of happy children,
> Knew a race had come to settle.
> Where for years they fished and hunted.
> So my people ran and gathered,
> Many paints and many brushes,
> tinted all the sky with colours,
> Clothed the hills with fresh green garments,
> Welcomed those who came with sunbeams,
> That reflected from the waters,
> Wished them well for health abounded,
> in this valley by the mountain,
> sent me here to give their message,
> sent me here to tell its meaning.[48]

The little girl who had sung 'Hiawatha' to a pre-World War I shipboard audience years before had clearly internalized not only the historical resonance of Longfellow's poetics, but also the nuts and bolts of his prosody. She had also, clearly, found it expedient to re-enact a reading of indigenous history in which Aboriginal people colluded in and accepted graciously the settler appropriation of their lands and the replacement of their traditional way of life by Euro-North American agricultural and industrial economies.

Yet these performances were paralleled by others in which Princess White Deer and her father advocated for the redress of that dispossession.

A photograph taken in the same year as the Lake Mohawk resort opening, 1937, shows her wearing the same Plains Indian beaded hide dress (identified in a newspaper report as 'the costume of the women of the Mohawk nation'[49]), she donned to participate in settler historical pageantry when she went to the White House to present an invitation to President Roosevelt to attend a Great Council of the Six Nations held at the Mohawk reserve of St Regis (Akwesasne) (Figure 10.3). The text of the invitation makes clear the sovereigntist political message underlying this commemorative celebration:

> Esteemed Sir: We, the Hodeinonkseonin, The League of the Iroquois, in the name of Kayanerengo (the Law of Great Peace), whose strands of eternal record have given the torch to Western

Figure 10.3 Princess White Deer at the White House

> civilization, extend unto you the heart's right hand, on the occasion of the ceremonial celebration of the greatest liberty conceived by man. Six hundred years ago We-Hin-Oh-Kwa gave to the world a new birth of matchless freedom, the undying principle of which a modern sick world needs.
>
> In the lands of the Mohawk Nation, on the St. Lawrence, at the St. Regis Reservation, on July 5th, 1937, beginning at 2 p.m., there will assemble for the first time since before the Whirl-Wind (the Revolution) the Grand Council of the ancient Iroquois Confederacy. We will be highly honoured by your presence on this historic occasion. [signed] George E. Thomas, Head Chief.[50]

One newspaper reported that 'the President gave Princess White Deer fifteen minutes of his time and indicated not only much interest in the ceremonies of 5 July 1937 but also in the progress and well being of the Six Nations which he promised to promote'.[51]

Although the Roosevelt years saw an important liberalization of the Bureau of Indian Affairs, most notably in Secretary John Collier's support for the protection of reserve lands and the preservation of Indian traditions, this shift was focused on the cultural, rather than the political life of Native American communities. That reality is reflected in the lack of response to the invitation on the part of the many political leaders, including the members of the President's cabinet and the Canadian ambassador, to whom White Deer also delivered invitations. None of these officials attended the meeting at which the Mohawk staged a formal adoption ceremony for an English pacifist named Charles David who had promoted the Hodenosaunee Confederacy as a model of organizations like the League of Nations. Both Esther Deer and her father participated in the ceremonies, which were reported in the the *Massena Observer*:

> Chief Deer strode onto the platform accompanied by two venerable Indians, all clad in full feathered regalia. Later his daughters, Running Deer and Princess White Deer, also in costume, joined the platform crowds.
> Their Indian garb contrasted with the heavy use of lipstick and mascara to combine the old and the new.
> In a colorful finale Davis was invested as a member of [the] Iroquois and given the name of Safe Keeper of the Council Fire as he was admitted to the Snipe Clan.

> Chief Deer got out the peace pipe and each puffed on upon it, after which the Indian draped a gaudy blanket over Davis's shoulder and put on the feather head-dress ...
> Princess White Deer sang Indian songs, and Running Deer gave an original blanket dance.
> The gathering was partly adjourned when Chief Deer again stepped up to the mike and reminded his audience how he had helped conquer the Nile at Khartoum.[52]

The mixture of popular entertainment and serious purpose that emerges from this description seems a fitting final act in Princess White Deer's public life, which was coming to a close as she approached her fiftieth birthday and as the world stood poised on the brink of another world war. The documentary evidence of her active participation as both a professional entertainer and a political activist contained in her scrapbooks ends at the beginning of World War II. The last reference in her papers to a public intervention is a 1942 letter from the governor of New York agreeing to her request to proclaim 26 September as American Indian Day.[53] Esther Deer lived on in New York for another half century, dying in 1992 at the age of 100. During her visits to her Canadian relatives at Kahnawake, she shared her memories of a long and successful show business career, and she left her mementos to her great niece Sylvia Trudeau, who later deposited them at the Kanien'kehaka Onkwawén: na Raotitiohkwa.

Playing on the Indian Princess

This account, however brief, provides abundant evidence both of Esther Deer's enactments of the quintessential Indian Princess and of her exploitation of the doubleness inherent in the virgin/temptress dichotomy that has been identified by numerous authors as intrinsic to its power.[54] The dramatic impact of many of her Indian acts was intensified by intentional juxtapositions of the modalities of the Princess persona with other 'commodities' of Indianness identified by Goldie, in particular Indian Spirituality and the Indian Warrior.[55] In addition, as we have seen, in other dances Esther Deer also played with signs of blackness, and her manipulation of the signs and symbols of Indianness and blackness exploited a specifically 1920s phase of primitivist discourse. One of the clearest signs of the primitivist framing of reenactments during the early twentieth century is a 1921 photograph in which Princess White Deer holds a Hodenosaunee *Gagoh:sah* or False Face mask next to her face

(Figure 10.4). Her gesture strikingly prefigures the famous photograph *Noir et Blanche*, in which a white woman holds a black Baule mask from Africa next to her face which the surrealist photographer Man Ray would make five years later. Esther Deer's juxtaposition of a 'savage' mask with her own very modern (and pale) face plays with the notion of the Indian as simultaneously strange and familiar, both claiming and calling into question her own identity as Other.

These images, created by an Aboriginal and a European, are perhaps the most graphic demonstrations of the potential for ironic unmasking contained in the dialogics of hybridity. In constructing these images Esther Deer performed not as a pawn on a chessboard moved by a ghostly semiotician or in 'someone else's story', but as a subject who

Figure 10.4 Princess White Deer with mask

was deliberately destabilizing the tropes of race and ethnicity. By constantly alternating and rearranging the signs of Indianness, of blackness, and of femininity, she dislodged them from their accustomed textual locations to explore the possibilities and the realities of doubleness. Her rapid changes of costume and of persona from 1920s flapper, to demure American patriotic heroine, to sexy Indian princess, were transformations characteristic of the kind of tricksterish play that calls standard icons into question. She made it impossible for her audiences *not* to know that the Indian stereotypes were as much (dis)guises as the other, more patently fictional images she adopted. Behind the glitzy Broadway images, too, we also catch glimpses of other doubled images that preceded them– the before and after representations of the savage and the civilized Indian so beloved both by mid-nineteenth-century artists like Charles Bird King and George Catlin and by the later wardens of Indian boarding and residential schools.

When Esther Deer later met President Roosevelt to present the invitation of the Hodenosaunee Confederacy Chiefs, her choice of garb was a traditional beaded hide Plains Indian dress that tacitly affirmed the survival of Indianness in the face of assimilationist expectations. Although non-Natives read the wearing of this outfit as a nostalgic reference to the past that confirmed their historical consciousness as the legitimate successors to the Indians, for Aboriginal people the donning of such clothing signified a continuity reaching from the past into the future. The evidence leads as convincingly to a reading of the displays of Indian artifacts and clothing worn in reenactments of historical events as oppositional gestures intended to counter notions of the vanishing Indian as it does to a reading of complicity with dominant tropes. In either case, the wearing of Plains feather bonnets and beaded clothing met audience expectations formed by dominant pan-Indian conventions and constructs of authenticity. For performers like Esther Deer, the dominant culture's identification of her as a 'REAL American' offered her not only the opportunity to further her own career, but also to enhance the efficacy of her work for Indian causes.

Robert Young has argued for the importance of Bakhtin's dialogic theory in the analysis of the hybrid forms that are produced by colonial interactions. 'Bakhtin's doubled form of hybridity', he writes, 'offers a particularly significant dialectical model for cultural interaction: an organic hybridity, which will tend towards fusion, in conflict with intentional hybridity, which enables a contestatory activity, a politicized setting of cultural differences against each other dialogically'.[56] This 'intentional hybridity', Young continues, both brings together and 'maintains a separation', permitting 'one voice to unmask another ... within a single discourse'.[57] The doubleness and ironic unmasking perfectly describe the Indian acts devised by entertainers like Princess White Deer. Rather than accepting the unidimensionality implied by true synthesis– the hybridity of fusion and assimilation– their performances exploited the contrasts inherent in their split cultural position between two worlds by throwing into relief the contrast between the rigidity of the imposed stereotypes and the fluidity of complex, modern Native identities.

The reenactment of Indianness in popular entertainment was a particularly potent form of intervention in mainstream discourses of Indianness because the spatiality and temporality of performance forms was ideally suited to the enacting of doubleness. Because they were so baldly fictive, historical reenactments, which were part of the stock in trade of popular performance, helped to erode and destabilize the stereotypes of Indianness. Princess White Deer's performances, like those of other contemporary entertainers, worked partly through her audience's pleasurable recognition in and comfortable familiarity with the known– the 'old Indian trick', Lenore Keeshig Tobias describes in her poem 'Goodbye, Wild Indian':

> I know – old Indian trick
> Just let them think
> what they want.
> They always do anyway.
> That's their problem, aaay.[58]

Yet at the same time, White Deer's rapid changes of costume, persona, music, and dance also called forth the audience's collusion in and affirmation of her co-modernity. The process worked to complicate and break down the two-dimensionality of the stereotype and to challenge the paradigm of authenticity at its heart. Ultimately the shared acknowledgement that Indians were not pre- but co-moderns would help to challenge the assumptions which had shaped not only the reenactments of Indian entertainers, but also the Indian Acts of parliaments and legislatures.

Notes

1. (2007) *The Double Entendre of Re-enactment: An Interactive Program with Gerald McMaster* (Toronto: V tape/AGO), p. 24.
2. An earlier version of this paper was published in J. C. H. King and C. F. Feest (eds) (2007) *Three Centuries of Woodlands Art* (Altenstadt, Germany: ZFK Press).

We have revised the paper to bring Esther Deer's career into sharper focus as an example of historical re-enactment. Details of Deer's career are given at greater length in the earlier version. Research for this paper was carried out between 1992 and 1995. We are grateful to Esther Deer's great niece, Sylvia Trudeau and to Kanatakta, Martin Loft, and Anna Mae Rice of the Kanien'kehaka Onkwawén:na Raotitiohkwa for their critical contributions to and facilitation of this research.

3. The Mohawk are one of the six nations of the Hodenosaunee confederacy, today living primarily in New York State, Ontario, and Quebec. 'Hodenosaunee', or 'People of the Longhouse' is the name preferred today to the earlier 'Iroquois'.
4. Franz Boas, for example, refused to invite Kahnawake entertainers to perform at the official Indian village of the Chicago Columbian World's Fair in 1893 because he regarded them as too acculturated, and no examples of the beaded garments worn by Hodenosaunee entertainers was knowingly collected by him or by the many ethnologists he trained to collect material culture for North American anthropology museums. See D. Blanchard (1984) 'For Your Entertainment Pleasure – Princess White Deer and Chief Running Deer – Last "Hereditary" Chief of the Mohawk: Northern Mohawk Rodeos and Showmanship', *Journal of Canadian Culture*, 1:2, pp. 99–116.
5. On assimilationist doctrines in the US, where Princess White Deer's career unfolded, see Frederick Hoxie (1997) *A Final Question: The Campaign to Assimilate the Indians, 1880–1920* (New York: Cambridge University Press). On the trope of the vanishing Indian, see B. Dippie (1982) *The Vanishing American: White Attitudes and U. S. Indian Policy* (Middletown, CT: Wesleyan University Press).
6. T. Goldie (1989) *Fear and Temptation: The Image of the Indigene in Canadian, Australian, and New Zealand Literatures* (Montreal: McGill-Queen's University Press), p. 10.
7. D. Francis (1992) *The Imaginary Indian: The Image of the Indian in Canadian Culture* (Vancouver, BC: Arsenal Pulp Press), p. 103.
8. D. Doxtator (1988) *Fluffs and Feathers: An Exhibit on the Symbols of Indianness, A Resource Guide* (Brantford, Ont.: Woodland Cultural Centre).
9. L. Keeshig-Tobias (2005) 'For<e>ward', in U. Lischke and D. T. McNab (eds) *Walking a Tightrope: Aboriginal People and their Representations* (Waterloo, Ont.: Wilfred Laurier Press), p. xvi.
10. *Double Entendre of Re-enactment*, p. 8.
11. Ibid.
12. H. Bhabha (1994) *The Location of Culture* (New York: Routledge), p. 85.
13. The unpublished diaries are in a private collection but are quoted extensively in B. McBride (1995) *Molly Spotted Elk: A Penobscot in Paris* (Norman, OK: University of Oklahoma Press).
14. McBride, *Molly Spotted Elk*, p. 90.
15. Ibid., p. 92.
16. Ibid., p. 136.
17. P. J. Deloria (1998) *Playing Indian* (New Haven, CT: Yale University Press), p. 122.
18. Blanchard locates the origin of Mohawk involvement in professional entertainment in the displays staged for the Prince of Wales during his 1860 visit

to Canada and to the invitation to visit England extended to Mohawk entertainers as a result of this visit. See 'For Your Entertainment Pleasure', p. 99.

19. James Deer also performed a strong man act, and was later certified to have lifted a horse with his teeth. See letter to Jas D. Deer from Walter L. Main dated 18 May 1938, Andover, Ohio. In *Esther Deer Papers*, Kanien'kehaka Onkwawén:na Raotitiohkwa (hereafter abbreviated as KOR, *EDP*).
20. John Deere, age 22 years, and James Deere, age 18 years, 'of Caughnawaga' appear as numbers 32 and 34 respectively on a list of 382 foremen and boatmen on the expedition. See C. P. Stacey (1959) *Records of the Nile Voyageurs, 1884–5: The Canadian Voyageur Contingent in the Gordon Relief Expedition* (Toronto: Champlain Society), p. 257.
21. 'The Deer Brothers Famous Champion Indian Trick Riders of the World: Assisted by the Equestrienne Wonder Gerogie Deer', KOR, *EDP*. Like this flyer, many of the programmes, newspaper clippings and photographs in the Princess White Deer collection lack dates. In our references to this archive we have included titles and names of newspapers or other details where they exist. We have also amplified this research by cross-referencing the other sources within the archive, but both the chronology and citation of sources should be regarded as provisional and will undoubtedly be refined by further research.
22. See http://www.boondocsnet.com/expos/wfe_1901_Ocg_midway.html
23. Undated newspaper clipping, KOR, *EDP*.
24. According to the programme, 'Indians of the Past' was performed in April 1908 at Edinburgh's Alhambra Theatre of Varieties. The programme also noted James and John Deer's role in the Wolseley expedition to relieve General Gordon, KOR, *EDP*.
25. Clipping from an unknown Buffalo newspaper, KOR, *EDP*.
26. See http://www.boondocsnet.com/expos/wfe_1901_Ocg_midway.html, as cited in note 22.
27. For Keith and Albee, see A. Slide (1994) *The Encyclopedia of Vaudeville* (Westport, CT: Greenwood Press), pp. 5–7; and see also A. F. McLean (1965) *American Vaudeville as Ritual* (Lexington, KY: University of Kentucky Press), p. 195.
28. Untitled newspaper clipping, KOR, *EDP*.
29. *The Savannah Press* (9 February 1918). The programme lists White Deer's act as 'The Original American Girl of the Mohawks in a Dancing and Scenic Act ...'. She visited Liberty Theatres in army camps in South Carolina, Georgia, Alabama, Arkansas, Iowa, Wisconsin, Michigan, Ohio, and Kentucky, KOR, *EDP*.
30. *The Scranton* [Pennsylvania] *Press* (n.d.), KOR, *EDP*.
31. KOR, *EDP*.
32. Ibid.
33. 'Hitchy Koo 1919 to Open Next Monday: Raymond Hitchcock will Present New Revue at Liberty Theatre. It's Mostly a Girl Show'; and review of Hitchy Koo 1919 at the National Theatre, KOR, *EDP*. According to Slide, the Hitchy Koo shows, which included comedy sketches and songs, were produced from 1917 to 1920 by Raymond Hitchcok, a monologist who also starred in the shows. See Slide, *Encyclopedia of Vaudeville*, pp. 241–2.
34. Article from an unidentified newspaper. It goes on to state that, preparing for the Indian role in 'Hitchy Koo', Esther Deer returned to Kahnawake to

'secure two full-blooded Indian maidens to appear with her. The names of the recruits are Miss Beaue is [*sic*] and Miss Skye', KOR, *EDP*.

35. Review of Hitchy-Koo 1919 at the National Theater, KOR, *EDP*.
36. 'Princess in "Tip Top"', *New York Tribune* (6 October 1920), KOR, *EDP*.
37. *New York Variety* (21 January 1925).
38. Unspecified publicity material, KOR, *EDP*.
39. The *Morning Telegraph* (7 March 1925) reported that 'Princess White Deer ... whose Indian revue, "From Wigwam to White Lights", was a sensational success at the Hippodrome and Eighty-First Street Theater, has been ordered to Florida by her physicians following a nervous breakdown due to overwork in preparing her elborate [*sic*] act for Keith-Albee Vaudeville ... While in Florida, she will study the last of the Seminoles in their island hunts [*sic*] in the Everglades', in KOR, *EDP*.
40. Untitled newspaper clipping, KOR, *EDP*.
41. *New York Theater Advertising Medium* (n.d.), p 44, KOR, *EDP*.
42. 'Mary Eaton Rises to Stardom with New Musical Play', unidentified New York newspaper (24 March 1927), KOR, *EDP*.
43. Reported in a letter from Mr Walter G. Kingsley, written on the letterhead of the New Amsterdam Theater to Mr Beth Beri, Paris, France (13 August 1928), KOR/ *EDP*.
44. Letter from Robert Redmond for the Committee for the National Tribute to General John J. Pershing (26 March 1925), KOR, *EDP*.
45. Unspecified, publicity material, KOR, *EDP*.
46. 'Notice', *The Boston Globe* (8 May 1927), KOR, *EDP*.
47. *The Daily Reporter* (Wednesday, 9 August 1933) and *The Daily Item*, Port Chester, NY (Tuesday, 8 August 1922), KOR, *EDP*.
48. 'Lake Mohawk to Open Sunday: Celebration Planned for Christening Development at Saprta: Indian Princess to Attend: Daughter of Head of Mohawk Tribe to Be a Feature – Land Filled with History', KOR, *EDP*.
49. Unidentified newspaper clipping, KOR, *EDP*.
50. Invitation from George E. Thomas to President Roosevelt, KOR, *EDP*.
51. The photograph accompanied a 1937 newspaper report in an unidentified Washington D.C. newspaper entitled 'Indian Princess Here to Call on President: Princess ... Special Ambassador ... Will Invite Roosevelt to Grand Council of Six Nations'. It listed other invited guests as including all the major political leaders in Europe and North America, the major newspaper owners, military leaders, leaders of major national associations and 'over 2,000 leading men and women', in KOR, *EDP*.
52. *Massena Observer* (n.d.), KOR, *EDP*.
53. Letter from Herbert H. Lehman, Governor, State of New York to Princess White Deer (7 October 1942), KOR, *EDP*.
54. See for example, Goldie, *Fear and Temptation*. Goldie's formulation has been much cited in the literature. For indigenous critiques of these stereotypes of indigenous sexuality, see D. H. Taylor (ed.) (2008) *Me Sexy: An Exploration of Native Sex and Sexuality* (Vancouver: Douglas and McIntyre).
55. Goldie, *Fear and Temptation*, pp. 15–17.
56. R. Young (1995) *Colonial Desire: Hybridity in Theory, Culture, and Race* (London: Routledge), p. 22.
57. Ibid.
58. 'Goodbye, Wild Indian' in Lischke and McNab (eds), *Walking a Tightrope*, p. 4.

11
Mimic Toil: Eighteenth-Century Preconditions for the Modern Historical Reenactment

Simon During

Like other genres that have arrived on the historical scene relatively recently, the historical reenactment is a more complex cultural form than its status would lead us to suppose. Indeed, the category 'historical reenactment' covers a variety of subgenres including: (1) the organized recreational imitation of a historical event by hobbyists; (2) the reproduction of a historical genre (e.g., the medieval tournament) or situation (e.g., a nineteenth-century mining town), often in the interests of tourism; and (3) the repetition of a historical event for the media, usually television. None of these subgenres in their present form existed in the eighteenth century, which is somewhat puzzling since so many of their constitutive elements already existed. It is almost as if, looking backward, the period is straining toward the reenactment as we know it. What exactly, we might ask, is preventing its emergence?

But this teasing counterfactual is not the question that this essay mainly addresses. I am more concerned with listing and examining the constitutive elements of the not-yet-existent historical reenactment in the eighteenth century on the grounds that certain aspects of the period can take on a new look – can reveal new patterns, new interactions, new problems – when viewed in the light of historical reenactments to come. Such a project obviously does not belong to positive historiography. Yet it is not quite a form of naïve presentism either, since, instead of representing the past as if it anticipated the future that actually unfolded, it describes a set of relations and interactions that only exist in connection to the future (but perhaps contingent) arrival of historical reenactments. For want of a better term I will call this set an 'assemblage', despite its not being intentionally produced, indeed despite there being at the time no sense of its possibility. It is as if historical reenactments, existing in the eighteenth century as a spirit of the future hovering irresolutely over the social scene, can be transformed for that future by their phantasmal presence. Or to put it another way: the future in which historical reenactments become a characteristic form of popular, recreational public memory is not considered here as a fulfillment of past potentials but as a suggestive absence that allows us to problematize that past in quite specific ways.

The prehistory of historical reenactments welcomes such an approach because it involves elements across a wide and dispersed set of social fields, processes, and levels. Indeed this dispersion is both the mark of reenactment's complexity as a cultural form and a sign of what I will call its deinstitutionalization – its falling outside the traditional sites and structures of mimesis and public memory. As soon as we inquire into where might we find the preconditions of historical reenactment, we are drawn toward (just to name some of the most important sectors) popular entertainments, private masquerades and hobbies, fictions, civil ceremonies and commemorations, historiography, cultural/literary theory, and religious rituals. And this is not to consider conscious repetitions of earlier situations or figures: Robespierre as a re-enactor of Brutus, for instance, or the expansion/reinvention of the Order of the Garter in George III's reign in order to reanimate ancient ritual and honor.[1] Of course, as will be apparent, I am already overreaching myself since the first problem, that the effort to describe the prehistory of contemporary historical reenactment encounters, is to ascertain whether precisely these – and not other – moments or forms, can be thought of as its constitutive alphabet. We cannot be sure, of course, and in that sense this essay belongs to speculative history. Nor can precise relationships between the various moments and forms be read off the historical record with the degree of transparency that historians conventionally claim for themselves. But at the same time this speculative uncertainty enables a certain inventiveness: as I say, it allows us imaginatively to hazard unexpected and not necessarily fulfilled patterns within eighteenth-century society and culture.

One way to begin this task (into which I can only take a few steps here) is to segment the concept of the historical reenactment not into its subgenres but into four of its basic conceptual components. We can understand historical reenactments as syntheses of: (1) practices of commemoration, (2) practices of mimesis or imitation, (3) expressive collective performances, and (4) specific relations to particular historical events (or genres of event) in which the present is not thought of as a typological repetition of the (Biblical) past under Providence (as it was for much Protestantism), or as a falling off from a classical ideal (as it was for forms of neoclassicism), or as a store of lessons for public life (as it was for historiography à la Bolingbroke for instance), or even as a stage in incomplete human progress (as it was for certain enlightened intellectuals).

These various components of reenactment's prehistory, while they exist across different settings and institutions, are not completely independent from one another either: they can be joined within more general processes working on them all. Of such processes two are immediately relevant. First is the (gradual and spotty) historicization of everyday life, by which I mean the sense that the conditions of ordinary life were under systematic and rapid transformation, and that previous generations had inhabited a significantly different world than current ones. (As Mark Phillips has argued, this sense is closely connected to the increasing emphasis on 'manners' and biography within various different modes of history writing.)[2] This sense becomes much stronger in the last two decades in the century and is only commonplace in Britain after the wars against revolutionary France: it is a ground for Sir Walter Scott's renovation of the historical novel, for instance, or, to give another example, it is a ground for the difference between James Boswell's *Life of Johnson* (1791) and John Smith's *Nollekens and His Times* (1828), the second of which takes the Boswell biography as its model but has a sense lacking in the first that manners and material culture had altered almost beyond recognition over a generation. And it is the historicization of everyday life that stimulates the view that the past requires forms of empathy to be understood. As Thomas Warton put it in his *Observations on Spenser's Fairie Queene* (1754), a well-known example of this turn to empathy:

> In reading the words of an author who lived in a remote age, it is necessary, that we should look back upon the customs and manners which prevailed in his age; that we should place ourselves in his situation, and circumstances; that so we may be the better enabled to judge and discern how his turn of thinking, and manner of composing were biass'd, influenc'd, and, as it were, tinctur'd, by the very familiar and reigning appearances, which are utterly different from those with which we are at present surrounded.[3]

The second general process, linking the various components of historical reenactments, to which I wish to draw attention is the visualization and concretization of the past. By this I mean not just the increased emphasis upon seeing the past (taking a spectatorial attitude toward it that may also involve flows of sympathy toward it), but the accelerated dissemination of detailed images of what the past authentically looked like or was imagined authentically to look like.[4] This process also involved a disposition to live among mementos, images, and traces of the past: take the habit of inserting old or imitation stained glass windows into houses and churches from the 1750s on, or the fashion for furnishing private libraries and public rooms with busts of great men from the past that made up much of the livelihood of fashionable mid-century sculptors such as John Michael Rysbrack.

***

I want to begin my brief description of certain components of the not-yet-existent historical reenactment with my most abstract and remote instance, that of cultural theory and particularly the theory of mimesis itself. My example here is Richard Hurd's preface to his 1751 edition of Horace's *Epistola Augustum*, entitled 'A Discourse on Poetical Imitation'. I have chosen Hurd's essay not so much because it is an important and neglected contribution to criticism (although it is), but because it lies close to the centre of Georgian intellectual life. The essay was written when Hurd was relatively young; he would later become George III's favorite prelate (as Bishop of Worcester he was appointed tutor to the Prince of Wales and turned down the Archbishopric of Canterbury) and a supporter of North's administration. At the beginning of his career, he was patronized by William Warburton, the influential Whig ecclesiastic, and was friendly with William Mason, the parson poet. Later he became something of an oracle for younger Anglican scholars/careerists such as Thomas Percy. I have also chosen to concentrate on Hurd because his *Essay on Chivalry and Romance* (1764) is often and rightly regarded as a key document in the revival of chivalric Gothicism, and Gothicism, especially where it reanimated chivalric ideas, is closely bound to the prehistory of historical reenactment, as we will see.

In effect Hurd, in his non-theological writings, can be regarded as harnessing Montesquieu's quasi-sociological (or 'philosophic') historico-geographical working through of the theory of uniform human nature (itself dependent on Locke's metaphysics) to a Whiggish agenda, so as to set it against the neoclassicism traditionally associated with the Patriot opposition, but also to align it to the Edmund Spenser/Elizabethan revival mainly associated with Tories and Patriots.[5] Hurd took up this agenda against Shaftesbury's cosmopolitanism and celebration of oligarchic polite culture, as well as against the Shaftesbury/Hutchesonian understanding of the moral sense (that is the disposition to benevolent sympathy) as innate. It was an innovative and risky project since it marginalized Lockean contract theory, did not strongly adhere to a progressive understanding of history, and stood outside sentimentalism. But it possessed enormous appeal especially once George III had taken the throne: the evidence is the popularity in particular of the three-volume edition of Hurd's *Moral and Political Dialogues*: it was reprinted 18 times before the century ended.

In Hurd's preface to Horace's *Letters* this project takes shape as a vigorous attack on Aristotlean mimetic theory and, more generally, on the rhetorical concept of literature. A series of controversies placed the category of imitation under stress in the 1740s: Hurd was inspired in particular by a claim, much touted in the *Gentleman's Magazine*, that Milton was a plagiarist; by the account of well-respected deistic Cambridge don, Conyer Middleton, of certain Roman Catholic rites as repetitions of pagan rituals; and by William Mason's imitations of Chaucer, Spenser, and Milton in *Musaeus: A Monody to the Memory of Mr Pope, in Imitation of Milton's Lycidas* (1747), which, like the Lauder case, sparked a lively discussion in the *Gentleman's Magazine*. Hurd accepts that as readers we are 'often at a loss to discern the original from the copy', but that in such cases it is most likely that we are confronted with an original rather than a copy, human nature being the same in all times and all places (see Hurd's 1761 edition of Horace's *Horatii Flacci Epistola Ad Augustum*).[6] And it is more likely that different authors, even across different cultures (as we would say today), produce similar works more than they have copied one another – unless, of course, we have good evidence otherwise. (Hurd argues too that imitation is more likely when we face similar forms or styles than when we face similar content.) This argument has several interesting consequences. Most importantly it gives value simultaneously not just to what Hurd calls 'primary imitation' (that is the direct mimesis of nature or the expression of passions) but to originality. As it aims to demolish the emphasis on literary imitation as adherence to a model (secondary imitation, as Hurd calls it), the distance between original and imitation narrows. This is key since the closing of that gap is one of the underpinnings of the historical reenactment, which exists, at least ideally, at the point where imitation meets repetition. At the same time, it points away from what was to become the most popular form of mimesis in later eighteenth-century culture – fictional mimesis, which, of course, radically restructures the relation between the world and its representation. In this essay at least, Hurd is not especially interested in imitation that requires the willing suspension of disbelief, but such indifference matters little here since fictional mimesis belongs to the margins of the prehistory of historical reenactments insofar as fictions by definition do not represent actual events.

At the same time Hurd's theory opens the way for books like the one proposed by Thomas Percy in 1762 – an anthology to be called *Specimens of the Ancient Poetry of Different Nations*, which was to include Welsh odes, Macpherson's Ossian, poetry out of *Hau Kiou Choaan*, Percy's own *Five Pieces of Runic Poetry* and *The Song of Solomon*, together with specimens of East Indian, Peruvian, Lapland, Greenland, and Saxon poetry.[7] It is worth noting that Percy included a section of Hurd's essay in his *Miscellaneous Pieces Relating to the Chinese* (1762). Hurd's critique of classical imitation has the potential to bring the whole range of human communication and culture into the domain of valuable representation. All human creative endeavor becomes available for appreciation – which is one (non-Rousseauvian) conceptual underpinning for a welter of identifications with the 'primitive' by moderns, of which the cult of Ossian was the most popular expression, though one not endorsed by Hurd himself.8 This cult too belongs to the proto-historical reenactment assemblage, since it intensified and democratized the sense that contemporaries were interacting with historical figures, a sense whose strongest statement I know is the early nineteenth-century French painter Anne-Louis Girodet's ideologically complex canvas *L'Apothéise des héros français morts pour la Patrie pendant la guerre de la Liberté* (1801). In it, ghosts of Ossian's characters welcome and mirror Napoleon and other French heroes in an Elysium – a life-after-death – (which Girodet calls '*L'Élysée aérien*') whence the spirits of the dead repeatedly return to earth in an inspirational and amicable cohabitation with the living, a cohabitation that (at another level) both expresses and displaces the jumbling of the past and the present characteristic of the historicization of everyday life.[9]

When Hurd comes to discuss less theoretical topics, as in his sociological, rather Humean, account of the emergence of chivalry, the effect is to undermine the progressive view of history and to present history rather as a sequence of differences united by the uniformity of natural processes.[10] In his *Letters on Chivalry and Romance*, he sees chivalry as a consequence of the Christian war against Islam (and, following Warburton, romance as Arabic in origin), as well as an expression of the political and economic structures of European society at the time and in particular the almost sovereign independence of the feudal lords (Bolingbroke's 'Gothick Liberty').[11] In one of his most sophisticated texts, his 'Dialogue on Foreign Travel' (1788), a critique of the Grand Tour and a hesitant defense of the English universities, he renovates the Ciceronian philosophico-essayistic dialogue genre (his acknowledged modern models are Shaftesbury, Addison, and Berkeley) by having the discussion take place not between 'fictitious characters' as he calls them but between two well-known historical figures: Locke and Shaftesbury.[12] The fictional Shaftesbury defends not just travel as a form of education but cosmopolitan polish as a requisite for the English gentleman, whereas Locke speaks for a more localized and professionalized education, showing some disdain for French and Italian manners.[13] The discussion takes place soon after the 1697 Peace of Ryswyk and is elaborated in terms that are both somewhat different from and the same as those pertaining to the 1760s, especially in relation to the debate over Englishness that took place as soon as George III began to appropriate Patriot politics and call upon Francophobia in the midst of the Seven Years War (1757–63).[14] To somewhat similar ends in his 'Dialogue on the Age of Queen Elizabeth', Hurd stages a debate between the Tory Arbuthnot and the Whig Addison, which pits Arbuthnot's nostalgia for chivalry against Addison's insistence on modern polish and liberty, and effectively joins Arbuthnot in affirming the Elizabethan aristocratic tournament as an expression of gentlemanly virtue untarnished by modern luxury partly on the grounds that Addison's insistence on viewing the customs of the period as problems of government rather than as manners pure and simple is misplaced. In bringing Addison, Arbuthnot, Shaftesbury, and Locke back to life in this manner, Hurd himself is involved in reenactment and revivification even if only in the sensorally deprived medium of writing.

***

The second element of the assemblage brought into play by prospective historical reenactment also belongs to literature rather than to performance or spectacle. In Laurence Sterne's *Tristram Shandy*, Uncle Toby and his servant Corporal Trim re-enact military sieges of William III's nine-year war with France, in particular the siege of Namur at which Toby received a serious groin injury. This is an instructive episode in the prehistory of historical reenactments because it seems so closely to anticipate the genre, if only in fiction. Were we fully to grasp the meaning and point of Uncle Toby's hobbyhorse (as Sterne famously calls Toby's intense interest in acting out the siege), it is tempting to suppose that we might also grasp at least certain aspects of the relations underpinning protoreenactment culture, particularly insofar as they are related simultaneously to militarism and to leisure.

Sterne's novel is written from a different position in the Georgian literary field than Hurd's. Sterne was at war with Hurd's friend and patron, William Warburton, the Bishop of Gloucester, who was a dominant figure in the mid-century intellectual scene. (*Tristram Shandy* mocks Warburton, and early plans for the novel had him cast as a major satirical butt.)[15] Indeed Hurd too is on record as denigrating Sterne.[16] Among Anglican clergy intellectuals, Sterne is connected to a radical secularizing and sentimentalizing of latitudinarianism (that is the tolerant, anti-ritualistic forms of Anglicanism, closer to the Dissenting traditions), whereas Hurd stands for a transformative synthesis between old Tory patriot and oligarchic Whiggish ideologies. Nonetheless Hurd's recovery and sanctioning of originality, his analysis of the originality/imitation relation through the category of repetition, and his Lockean philosophic anticlassicism and anti-antiquarianism, as well as his revival of chivalry – are all shared in some form or other by Sterne.

Uncle Toby stands as an extreme case in this nexus because he does indeed belong to modern chivalry. His most insisted-upon character traits – modesty/chastity and bravery/militarism – are exactly what define chivalry in Hurd's *Essay*, although, in Toby, chivalry joins the genealogy of sentimentalism, a move of which there is no indication in Hurd.

In this context the central question becomes: why are Toby and Trim such obsessive re-enactors of old military sieges? The explanation that the novel itself puts forward at the beginning of its second volume is that Toby, while laid up in hospital with his rather mysterious groin injury (we are first led to believe that his genitals have been destroyed, although in later volumes that turns out not to be so), finds it impossible to describe in words the circumstances under which he received his injury. The reason for this is that the fortifications and siege apparatus at Namur were unprecedentedly tactically and spatially complex. So he turns to visual aids. It is his servant Trim's idea – an original, not to say an eccentric, idea – to substantially elaborate on these visual props by building a miniature version of the fortification on a bowling lawn belonging to Toby's property back home. This is what they set out – obsessively – to do, even though the original purpose of the visualization of the Namur siege (to make it easier for Toby to communicate to others) is now overturned, since as re-enactors Toby and Trim insist on privacy.

Toby and Trim's reenactments acquire meaning across a number of registers. First let us recall that Namur in particular was a military event of exceptional importance within British nation building. It was one of William III's major successes in the Nine Years War against France (1688–97), allowing the Protestant alliance to stem French control of the Low Countries. From a technical point of view, it is deemed a military and diplomatic milestone.[17] William himself was of course a hero of Whiggism given his role in bringing about the English constitution established in 1688–9; France was once again Britain's military antagonist in the Seven Years War, which was being fought as Sterne was writing (victory would seal Britain's global power for a century and a half to come); and, as we have already seen from our brief discussion of Hurd's 'Dialogue on Foreign Travel', cultural and political differences between Britain and France were at the center of public debate. Indeed Hurd's return to the period around 1700 in the 'Dialogue' (a few years after the first volumes of *Tristram Shandy* appeared), in terms that can be described as a soft form of textual reenactment, shares something of the spirit of the labors that Toby and Trim undertake on their provincial bowling green.

But what is disconcertingly at stake with Toby and Trim is the discrepancy between Namur's historical importance and resonance and Uncle Toby's extreme shyness and incompetence (not to say stupidity) in civil life. At a minimum his efforts at reenactment, not least as they become socially uncommunicative, can be seen as an attempt to participate in world historical events from the position of the private gentleman in retirement, where implicitly retirement has both its first sense – withdrawal from the civil sphere – and its more modern sense – the giving up of a profession after due service.

Given this failure to account for the form of life to which Toby turns in retirement, it is tempting for us today to read Toby's reenactment mania as a symptom of post-traumatic stress disorder: a compulsive repetition of a situation involving pain, shock, and loss, which is in the process of being worked through. It is a reading that might even be extrapolated to form the germ of a cultural theory of at least certain kinds of reenactments as post-traumatic events. It is true that, given both the indirectness and openness of Sterne's narrative technique and Toby's own lack of introspection and self-explanation, readers are invited to interpret – over interpret – his hobbyhorse. But the text itself does not even begin to engage in this kind of socio-psychological reading, going out of the way if anything to emphasize Toby's strange indifference to the consequences of the damage to his groin. Instead it is the satirical sentimentality with which he is presented that mobilizes a set of ambivalences and contradictions belonging to the 1760s: the relation between Protestant nationalism and European cosmopolitanism, chastity versus libertinage, social participation versus retirement, marriage (for men) versus bachelorhood, professionalism versus classical ruling-class *otium*, and the contradictions attached to the notion of modern chivalry at the point at which it is just beginning to underpin the gentlemanly ideal.[18] Toby's reenactments, which are both ridiculous and loveable, are designed to set these particular tensions into play.

***

My third element in the prehistory of historical reenactments comes from the world of performance almost a decade later than Sterne's *Tristram Shandy* and Hurd's *Essay on Chivalry and Romance*. In June 1773 George III reviewed navy ships at Spithead in a massively publicized event. Indeed the review was soon staged by all the licensed London playhouses: Samuel Foote's Haymarket, David Garrick's Drury Lane, and Covent Garden. In Drury Lane, a scene representing the occasion was added to the *Masque of Alfred*, a dramatic masque been produced in 1740 that was connected to the anti-Walpole opposition of the time.[19] The 1773 version, designed by Jacques de Loutherbourg and painted by Dominic Serres, is regarded as path breaking within the history of stage machinery and realism since it was one of the very first productions to use three-dimensional moving objects on tracks – in this case, models of navy ships.

The *Whitehall Evening Post* reviewed the production like this:

> This most singular exhibition is an incontestable proof of the rapid progress of the British arts. The general view is so critically exact, that one can hardly give human invention credit for the execution; and wonderful as it may appear in point of distance, perspective, &c. it is not chargeable with smallest impropriety.
>
> The view of Spithead and the fleet is taken from the saluting battery which we here see mounted with cannon. Every ship of the line is a beautiful perfect model, with rigging, &c. compleat, dressed with their proper suits of colours and carrying their regular number of guns; the Isle of Wight in the back ground forms a just and beautiful relief, the royal yacht is seen sailing into the harbour, under a salute of the battery, and the whole fleet. Numberless and various kinds of vessels are beheld under weight, with their sails full, making their different tacks, amongst which is readily distinguished the model of the beautiful cutter, belonging to the Duke of Richmond, remarkable for its blue and white striped sails. The deception of the sea is admirable. In short, the whole representation cannot but suffer from this, or any other attempt to describe it.

> While the public gratitude rewards the enterprising and noble spirit of the manager in this his wonder work, let it be at the same time be remembered, that the artist to whose abilities we are indebted for it, is the ingenious and well-known DOMINICK SERRES.[20]

This review can be connected to Toby and *Tristram Shandy*, and not just because it involves the military. At one level it is a spectacular representation of an actual event committed to maximum recognizability and authenticity – the *Whitehall Evening Post's* own review refers to the effectiveness of visual versus linguistic media for the reproduction of historical events. As was the case for Toby, it also involves a personal biography: Dominic Serres, a well-known public figure, had been a sailor, a prisoner of war, and a shipyard painter before setting himself up very successfully as a naval artist in London. But what distinguishes it from Toby and Trim's efforts on the bowling green is not just that it takes place in the public sphere or that it is a reproduction of a recent, almost contemporary, event, but that it has no status at all as a hobby-horse. This reproduction of the Spithead review, like those at the other theaters and like the paintings, models, and news reports that surround it, lie at the very heart of the nation's sense of itself.

As such it stands as one culmination of a long and varied history of staged or semi-staged mimetic and commemorative presentations of actual events. It is not my aim here even to begin to offer a conspectus of this history, which unfurls itself across a number of different media, but I will list a selection of instances across the century just to give its flavor. In 1704 a puppet show presented scenes from the Battle of Blenheim to London audiences, perhaps an early show by Martin Powell's troupe.[21] In 1711 Powell's troupe certainly staged a show advertised in the *Spectator* like this: 'The History of King Bladud, Founder of the Bath. The Figures being drest after the manner of the Ancient Britains. With the Walks, Groves, and Representation of the King's-Bath and new Pump-house. The Figures of Ladies and Gentlemen all moving in real Water'.[22] That same year Mrs Salmond's waxworks, which was primarily dedicated to works of no or doubtful historicity, included a piece featuring King Charles I on the Scaffold, which was to be a mainstay of her show through the century.[23] Charles I's life was also presented at Lincoln Inn's fields in 1736, in a drama avowedly modelled on Shakespeare's history plays but with 'characters dress'd in the manner of their time'. In 1727 George II's coronation ceremony was re-presented on the Drury Lane stage. Gilbert West's very successful oppositional *The Institution of the Garter* (1742) reproduced (along with much else) a supposedly authentic version of the first Order of the Garter ceremony dating back to the fourteenth century. This reenactment was revived in a more militaristic setting in 1771.

These instances begin to indicate that across the century non-fictional dramas based on real events (let us call them commemorative mimetic shows) were not only increasingly committed to authenticity, and that the increasingly sophisticated machinery of stage illusionism could be harnessed to that end, but that they were slanted toward occasions in which the nation's history and understanding of itself were at stake, and thus in particular toward military occasions. These events were often efforts of national self-glorification or attempts to mobilize national energies (even if residually divisive ones such as in the case of memories of Charles I), but they could be more than that since, in representing historical turning points, they also disseminated a sense of participation among the population, even among those not actually involved in the actions reproduced in cases where commemoration involved more or less contemporary happenings. And the possibility of mass participation would appear to have provided some of the energy through which mimesis was able to break out of traditional media such as the stage or the waxwork, a deinstitutionalization in which ordinary life itself could become its medium (viz. the name 'living history'),which it does in the case of Toby, or, somewhat differently, in historical reenactments proper.

***

There are signs that the war with the American colonies significantly further closed the gap between life and mimesis in terms relevant to the deinstitutionalization required by historical reenactments. I want to draw attention to three cases that point in this direction, each already well known to historical scholarship.

The first are the camps that were created to train and exercise the militia in Britain, most famously in Warley in 1778 and in Coxheath in 1779, and whose relation to the wider culture has recently been discussed by Stephen Conway and Gillian Russell.[24] Both camps were constructed during the invasion scare that followed the French alliance with the Americans, that is, after the war had become another struggle for global commercial dominance and prestige. As such it required what Conway has called 'a full mobilization of British resources'.[25] (Conway estimates that the British military participation rate in the American Revolutionary Wars for available men was between one in seven or eight compared to about one in five for the Napoleonic Wars and one in two for World War I.) Certainly the American war involved an unprecedented degree of volunteerism, even outside of Ireland, volunteerism's center, in part because the act of volunteering carried connotations of patriotic resistance to the Tory and Rockingham administrations.[26] Many volunteers joined without official encouragement or sanction, in part because of large-scale private fundraising efforts. And the camps, which housed both volunteers and regular troops, were figured by the Crown as emblems of the general population's unity and support for the war, which was, of course and especially before France became involved, much less than total. (Indeed one of the factors that projected this war much more deeply into national life than earlier wars was precisely its divisiveness.) Thus – to take one instance camp tents were represented in the background of Benjamin West's full-length portrait of the King, which was the showpiece of the 1780 Royal Academy's annual exhibition. It presented George in a military uniform, wearing the star and sash of the Royal Garter and holding a scroll of paper inscribed with the words, 'Plans of the Camps of Cox Heath, Warley, St Eden, [Ports]mouth, and Plymouth with a General Return of your Majesty's Forces in Great Britain August 18 1779'. (In light of our early encounter with the memorialization of the 1690s, it is worth noting that West also showed two representations of William III's military triumphs at the exhibition, namely *The Destruction of the French Fleet at La Hogue* and *The Battle of the Boyne*.)

Of course the camps were more than exercises in propaganda. The number of men they housed was unprecedented: over 15,000 in the case of Coxheath.[27] The camps were primarily designed to train the militia, and one important tool in that training was itself mimetic: the mock battle. These attracted large audiences, kept separate from the soldiers themselves by soldiers, as we can see in Philippe de Loutherbourg's 1779 canvas, *Warley Camp 1778: The Mock Attack*, now in the Royal Collection. The camps also quickly became tourist destinations: maps of their layout were printed; medals and coins were struck as souvenirs. London Society took the trip south to visit them, in part because the regiments were headed by aristocrats who welcomed their London friends. Intellectuals like Edmund Burke and Samuel Johnson also traveled down. Such tourist trips blurred the distinction between tourist and toured just because most visitors were hosted by friends or family in the camps. (Thus Samuel Johnson, who was visiting a friend, spent time in the tents of both officers and men and was shocked at the difference between the two classes.) The Duchess of Devonshire and her friends dressed in military uniform and formed a mock regiment of aristocratic women. Family hangers-on lived on the camps' outskirts.

This blurring was also intensified because the camps were commercial magnets: they attracted large numbers of sutlers, opportunistic traders and sex workers. It was in this liminal world that Richard Sheridan set his musical comedy of camp life, *The Camp* – basically an excuse for the presentation of de Loutherbourg and the Drury Lane artists' detailed stage set of the scene, but, in its self-referentiality, surprisingly Brechtian for all that. *The Camp*, set at Coxheath, is about a young women who disguises herself as a soldier in order to secure her lover's discharge, but it involves (self-referentially) a painter down from Drury Lane under instructions from de Loutherbourg to make sketches for the stage reproduction of the scene, and a group of soldier-struck market women, who work as prostitutes if the chance arises.

In sum then: these camps form part of the prehistory of reenactment assemblage because, once again, they break down the division between imitation and life – the mock battles are for real and not for real; they are both spectacles and serious occasions for training – and they muddy the division between audience and participant, because they draw mimesis into large-scale tourism, and, no less importantly, because they produce series of spin-off representations of themselves, many under the sign of authenticity. It is not that these preparations for war coincide with their mimetic representation, nor that they mobilize the whole population as would be the case for twentieth-century wars, but they are already pointed toward those eventualities. These are some of the conditions of possibility for large-scale reenactments, and it is no accident that in the US (not so much in the UK for obvious reasons) the American Revolutionary War is both a primary resource for historical reenactments and living history, organized around a version of that self-spectacularization that was characteristic of the camps to begin with.

My next example of a Revolutionary War proto-historical reenactment is both more famous and more directly related to modern reenactments. It is the so-called *Mischianza*, an elaborate entertainment produced in 1778 by British officers stationed in Philadelphia in order to mark the enforced return to Britain of their general, Sir William Howe. Funded by a select group of officers who contributed the enormous sum of over 3000 guineas for the evening's festivities, and attended by 400 ticketed guests, it involved a regatta, a mock medieval tilt and tournament 'according to the customs and ordinance of ancient chivalry', processions through newly constructed triumphal arches, completed by an extravagant Chinese-style fireworks show (constructed by the regiment's engineer) ending with an emblem of Fame spelling out the phrase '*Les Lauriers sont immortels*'); a supper (served by African slaves in Oriental costumes) and a ball.[28]

It is the medieval tournament that is of most interest to us among this festival of global exoticism. It was organized as a contest between two groups of knights, dressed in what was deemed to be the style current during the reign of Henry IV of France. These were the Knights of the Blended Rose (White Knights) and the Knights of the Burning Mountain (Black Knights), each Knight with their own device or visual motif and being accompanied by his own esquire. Each Knight was paired with a Lady, whose favor he fought for: each Lady was costumed in a Turkish turban and a diaphanous (presumably neoclassical) flowing gown (this being standard attire for women in the world of aristocratic private theatricals in the later 1780s).[29] The Knights challenged each other to various forms of combat, including a pistol duel, before saluting the Ladies, each group being judged equal and heading off to supper in a room 'festooned over with Flower knots'.

Clearly this is not a reenactment in the modern sense: it makes no serious claims to authenticity. Rather it is a hybridized mix of fashion and motifs. And it is another ideologically complex event, although it is important to note that it does not seem to be an act of persuasive communication to the local Philadelphians, many Quakers of course, even though the tournament's 'Ladies' were indeed local young women. Most immediately, it was a show of support for the British general and commander, William Howe, who was being removed for inaction and was to be replaced by the somewhat more professional Henry Clinton: as such it is a celebration of aristocratic amateurism and honor. It was also an act of conspicuous consumption, almost a potlatch, designed to intimidate Washington's army, starving at Valley Forge, only 18 miles to the north east (indeed the Americans fired a cannon at Philadelphia during the festivities). At another level the fashionableness and cosmopolitan air of the event advertized precisely the British officer's global connections and centrality.

The *Mischianza's* prime mover and designer was Major John André, whose tastes and interests it expresses. Unlike many of his fellow officers, André was not a member of the landed gentry, and had no great patron or connections to call upon. He came from a Lichfield trading family, and had worked in London commerce before joining the Army. More than that, he was an intellectual and aesthete, who, as a young man, had been part of the Lichfield circle that included Anna Seward, Erasmus Darwin, Brooke Boothby, Thomas Day, and Richard Lovell Edgeworth. Indeed he had been the official lover of Seward's honorary stepsister and love object, Honora Sneyd, although he was not rich enough to marry her: she became Edgeworth's second wife instead. He had also been very close to Anna Seward who was to write a popular 'Monody' to him (in the style of Hurd's friend William Mason) after his death. He spoke several European languages, was an amateur poet, painter and musician, especially interested in visual design, including the design of women's clothes.

André's reputation has been overshadowed by the manner of his death: he was hung by the Americans as a spy after having been caught behind enemy lines in possession of letters in which American general, Benedict Arnold, treated with the British to sell them the fort of West Port. But he brought to the *Mischianza* a flavor of the avant-garde, not to be expected among British Army officers. For the Lichfield group, among whom his tastes and cultural disposition had been formed, included some of the most fervent Rousseauvians in Britain, a connection that enables the mix of what we can broadly term Rousseaunism and Gothicism (along with Oriental exoticism) that characterizes the *Mischianza*. The Lichfield circle had joined in that theatricalization of elite leisure which was sweeping through Britain (let Brooke Boothby and Anna Seward's 1783 gothic pageant for Henry Fuseli, which involved impersonation of medieval knights, stand as an example) but the *Mischianza* tournament also contains a flavor of the *fêtes de la rose* that the Rousseauvian educationalist Comtesse de Genlis had been popularizing across France since the late 1760s. These were one of the earliest attempts to reanimate threatened rural festivals in the name of innocent, pre-modern collectivity. [30] In the light of the festival of the roses cult, the British army officers are laying claim to a certain chivalric inheritance and ethos through mimesis but they are doing so among flowers (and in particular among roses) that represent precisely a social ideal of naturalized innocence that will underpin romanticisms and revolutions to come. In terms of the theory of historical reenactment, this raises the question: to what degree can we suppose that reenactment is also entwined with reanimations of the organic grounds of social life, as filtered through the broad Rousseauvian critique of modernity?

***

My last element of the assemblage of proto-historical reenactments, although it concerns a portrait being painted, takes us back to Hurd's theory of imitation. The portrait is John Trumbull's *General George Washington at the Battle of Trenton* (1792). It's a full-length painting of a heroic Washington, represented in a gold tunic and trousers set in a war scene, caught 16 years after the event and not at all realistically, at the moment when Washington was inspired to conceive the tactical maneuver by which he cut off the British communication lines, and which led to the unexpected and decisive American victory the next day over the British and Hessian forces led by General Howe at Princeton. Scenes from the battle of Trenton are reproduced in the canvas's bottom left, overshadowed by Washington's heroic body. The portrait was commissioned from Trumbull by the city of Charleston, South Carolina who rejected it once they saw it, on the grounds that it did not reflect any particular association between Washington and South Carolina itself.

The portrait is also associated with John Trumbull's personal biography. Trumbull, from a Connecticut ruling-class family, had joined the revolutionary forces but, becoming disenchanted with the army bureaucracy, left to train as a painter in London at Benjamin West's studio. In London, he was arrested as a spy as pay-back for John André's execution. He was not executed though, and was soon released from jail, presumably because of West's influence with the Crown, and because pro-American sympathies were widespread among the Whig oligarchy. So Trumbull's relation to his portrait of Washington is personal in something like the way that Serres is related to his images of the Spithead review or even Toby's prodigies of reenactment are connected to his personal military past.

The painting contains a great deal of what Hurd called 'secondary imitation'. Washington's horse, which figures largely, is modeled on a horse from the Parthenon frieze, in the kind of neoclassical echo with heroicizing force that was commonplace in portraits of powerful public figures at the time. More disturbingly, or at any rate, more ironical, given West's relation to America's enemy George III, it also contains traces of Benjamin West's work. West's most famous contribution to painterly practice was his decision to paint his scene of General Wolfe's death during the Seven Years War (*The Death of General Wolfe*, 1770) simultaneously in the grand historical style and in modern dress. This was regarded as a breakthrough in the historicization of the present since it implied that the full weight of world history was being borne by contemporary figures, and it taught a lesson that the young American artists who trained under West were particularly quick to capitalize upon since they understood that history was being made in their home territory with exceptional rapidity and reach.[31] Trumbull's portrait of Washington extends the possibilities stimulated by West's painting of General Wolfe, but, more immediately, it also contains formal echoes of West's portrait of George III which I discussed above, and which of course presents a martial King against a contemporary military scene (in that case the training camp and the naval review). The ironies of the echo between West's glorification of George and Trumbull's glorification of George Washington should be obvious.

But Trumbull's painting is also very different from West's in that the relation between artist and sitter, between past and present, and between ideology and portrait convention interact through very different and very particular channels. After all, Trumbull is not quite painting the Washington whom he sees in front of him in his Philadelphia studio (for that Washington we can look at Gilbert Stuart's portraits of the period.) He is painting an imaginary Washington who exists in an indeterminate time between the Battle of Trenton and now, when he is, in fact, President of the new nation state. Indeed Trumbull is painting the elder Washington *becoming* the younger Washington, re-enacting his earlier military self. Washington is dressed just like he was at Trenton, and as Trumbull painted him, he and Washington (in Trumbull's own words) 'talked of the scene [before the Battle of Trenton], its dangers, its almost desperation. He *looked* the scene again, and I happily transferred to the canvass, the lofty expression of his animated countenance, the high resolve to conquer or to perish'.[32] It is impossible to imagine a European artist and head of state engaging each other in this way so as to capture a moment of living history. That requires a particular political constitution and ideology, a particular crux in the history of a nation, a particular relation between artist and hero – let's call it 'masculine history-making democracy'. And to return to Hurd, if Trumbull's portrait is formally imitative, its content is even more original than Hurd's theory can account for, precisely insofar as it provides the occasion for what is in effect a lived act of aesthetic imitation itself rather than a mimetic representation of the past.

****

It would be tempting to search for a general frame for all these moments in the assemblage that constitutes historical reenactment's prehistory, and then to connect them to a conjunction (the aftermath of the American Civil War?) when modern reenactments *really* begin. But to succumb to that temptation would be to miss this essay's limited aims. As I said at first, it is interested merely in allowing elements and patterns in eighteenth-century society and culture to suggest themselves, more or less imaginatively, when parsed in terms of reenactments to come so as to permit the spirit of future reenactments to remain precisely phantasmal, a hovering absence, animating uncertain connections of the kind with which we have been concerned here.

Notes

1. See L. Colley (February 1984) 'The Apotheosis of George III: Loyalty, Royalty and the British Nation 1760–1820', *Past and Present*, 102, pp. 94–129.
2. See M. Phillips (2000) *Society and Sentiment: Genres of Historical Writing in Britain, 1740–1820* (Princeton, NJ: Princeton University Press), p. 202ff.
3. T. Warton (1754) *Observations on the Faerie Queene of Spenser*, (London: R. J. Dodsley), p. 217.
4. For general material on some aspects of this, see D. Lowenthal (1985) *The Past is a Foreign Country* (Cambridge: Cambridge University Press), pp. 125–84.
5. For an excellent account of relations between the Walpolean opposition and the retrieval of renaissance and medieval culture, see C. Gerrard (1994) *The Patriot Opposition to Walpole: Politics, Poetry, and National Myth, 1725–1742* (Oxford: Oxford University Press).
6. Horace (1751) *Horatii Flacci Epistola Ad Augustum, with an English Commentary and Notes; to which is added A Discourse Concerning Poetical Imitation by R. Hurd and R. Bentley* (London: W. Thurlbourn), p. 109.
7. B. Davis (1989) *Thomas Percy: A Scholar-Cleric in the Age of Johnson* (Philadelphia: University of Pennsylvania Press), pp. 92–3.
8. Hurd was among the most acute critics of James Macpherson's claims, just because of the attention he paid to imitation. See S. Brewer (1995) *The Early Letters of Bishop Richard Hurd, 1739–1762* (Rochester, NY: Boydell & Brewer), p. 385.
9. See S. Bellenger (2005) *Girodet: 1767–1824* (Paris: Gallimard), p. 234ff.
10. For a discussion of the Scottish enlightenment account of chivalry, which mainly postdates Hurd and is not notably more sophisticated, see J. G. A. Pocock (1988) *Virtue, Commerce, and History: Essays on Political Thought and History, Chiefly in the Eighteenth Century* (Cambridge: Cambridge University Press), p. 198.
11. In Hurd's *Dialogue on the Age of Elizabeth* (in which he pits Arbuthnot's nostalgia for chivalry against Addison's Whiggish insistence on modern polish and liberty), the Elizabeth aristocratic tournament becomes an expression of gentlemanly virtue untarnished by the vicissitudes of modern luxury. Warburton's contributions to the historiography of romance are to be found, most easily, in a note to *Love's Labour Lost* in his edition of Shakespeare. See A. W. Evans (1932) *Warburton and the Warburtonians: A Study in Some Eighteenth-Century Controversies* (London: Oxford University Press), p. 120ff.
12. R. Hurd (1788) *Moral and Political Dialogues; with Letters on Chivalry and Romance*, vol. 1 (London: T. Badell), p. xxiii.
13. The inspiration for this dialogue is to be found in a letter by Warburton to him. See Brewer, *The Early Letters*, p. 251.
14. For a sustained discussion of Francophobia, see L. Colley (1992) *Britons: Forging the Nation, 1707–1837* (New Haven, CT: Yale University Press).
15. For Warburton and Sterne, see M. New (1982) 'Sterne, Warburton, and the Burden of Exuberant Wit', *Eighteenth-Century Studies*, 15, pp. 245–74 and A. H. Cash (1986) *Laurence Sterne: The Later Years* (London: Routledge), p. 69.
16. Brewer, *The Early Letters*, p. 368.
17. See J. Hoppit (2000) *A Land of Liberty? England 1689–1727* (Oxford: Clarendon Press), pp. 101–6.

18. See M. Girouard (1981) *The Return to Camelot: Chivalry and the English Gentleman* (New Haven, CT: Yale University Press).
19. R. G. Allen (1960) 'The Stage Spectacles of Philip James de Loutherbourg', (PhD dissertation, Yale University), pp. 124–27.
20. *Whitehall Evening Post* (9 October 1773).
21. It is interesting that George Speaight does not record this event in his standard *History of the English Puppet Theatre*, and he has the Powell troupe arriving in London from Bath in 1705.
22. *Spectator* (1711), p. 14.
23. R. Altick (1978) *The Shows of London* (Cambridge, MA: Belknap Press), p. 52.
24. See S. Conway (2000) *The British Isles and the War of American Independence* (Oxford: Oxford University Press), and G. Russell (1995) *The Theatres of War: Performance, Politics, and Society, 1793–1815* (Oxford: Oxford University Press).
25. Conway, *The British Isles and the War of American Independence*, p. 16.
26. See S. Conway (2004) '"Like the Irish"? Volunteer Corps and Volunteering' in J. Flavell and S. Conway (eds) *Britain and America Go to War: The Impact of War and Warfare in Anglo-America 1754–1815* (Gainesville, FL: University of Florida Press), pp. 143–72.
27. For a detailed description of the camps and everyday life inside them, see J. R. Western (1965) *The English Militia in the Eighteenth Century: The Story of a Political Issue, 1660–1802* (London: Routledge & Paul Kegan), p. 387 ff.
28. My information about this show is mainly drawn from John André's own account, which was printed in the *Gentleman's Magazine* and the *Annual Register*, available at freepages.history.rootsweb.com/~wcarr1/Lossing1/Chap36a.html. Also useful is Morris Bishop's article first published in *The American Heritage Magazine*, available at http://www.americanheritage.com/articles/magazine/ah/1974/5/1974_5_69.shtml.
29. For descriptions of the elite private theatricals of the day, and illustrations of costume, see S. Rosenfeld (1978) *Temples of Thespis: Some Private Theatres and Theatricals in England and Wales, 1700–1820* (London: Society for Theatre Research).
30. For the Fuseli pageant, see M. Myrone (2004) 'Gothic Romance and the Quixotic Hero: A Pageant for Henry Fuseli in 1783', available at http://www.tate.org.uk/research/tateresearch/tatepapers/04spring/gothic_paper.htm. For Madame de Genlis's rose festival, see S. C. Maza (1993) *Private Lives and Public Affairs: The Causes Célèbres of Prerevolutionary France* (Berkeley: University of California Press), p.73.
31. For a discussion of the context of the painting and the controversy, see H. von Erffa and A. Staley (1986) *The Paintings of Benjamin West* (New Haven, CT: Yale University Press), pp. 59–85.
32. J. Trumbull (1841) *Autobiography, Reminiscences and Letters of John Trumbull from 1756 to 1841* (New York: Wiley and Putnam), pp. 166–7.

12
Loutherbourg's Simulations: Reenactment and Realism in Late-Georgian Britain

Iain McCalman

On 28 June 2005 Queen Elizabeth formally reviewed 167 ships from Britain's Royal Navy fleet plus some 30 international naval vessels as they conducted a sail-past in the Solent at Spithead near Plymouth. Part of a year-long series of events to commemorate the two-hundredth anniversary of the Battle of Trafalgar, the review was a massive, popular pageant, pre-advertised in British tourist publicity for several years in advance, attended on the day by 300,000 holidaymakers, and watched on subsequent television broadcasts by millions all over the globe.[1] According to press and public relations releases, the Spithead naval review was part of an ancient and proud British naval tradition. Though informal fleet reviews were said to go back 600 years, the first formal Spithead Review had been undertaken in October 1772 before George III, and had been replicated with increasing pomp and circumstance by successive British monarchs ever since. In all the torrent of hype no one mentioned, however, that George III's Spithead Review could also boast another type of primacy: it had been the first ever visual simulation and mass-media reenactment of modern history.

The Spithead Review simulation was the creation of an Alsatian-born British-based artist, scene designer and spectacle inventor Philippe Jacques de Loutherbourg, and it was the first of a series of mechanical and/or pictorial reenactments of historical events – recent and remote – invented as byproducts of his theatrical repertoire. Popular success also later led him to develop his reenactments further into two different types of spectacle which flourished within the commercial entertainment world of late Georgian London – a moving picture show of the 1780s called the *Eidophusikon* and a series of large-scale panoramic battle paintings created for popular exhibition during the 1790s. This flowering of late-eighteenth century reenactment constitutes, I believe, a crucial foundational phase in the genre's long march towards modern-day cinema, television and museology. Despite the obvious differences between our entertainment media now and then, Philippe de Loutherbourg's mechanical reenactments anticipated a number of characteristics that we have come to associate with modern historical reenactments.

Firstly, they assumed several of the same generic forms that Jonathan Lamb sees as typical of today's reenactments – especially social and community pageants, intimate theatrical performances, and various modes of literalist realism.[2] Secondly, they quickly spread from London's traditional patent theatres to enter a new crop of Georgian popular commercial entertainment genres, most of them based on visual technologies. Thirdly,as in our own day,their popularity derived from adopting a popular aesthetic of mimetic 'realism', which, in extreme form, claimed to produce illusions indistinguishable from nature, including the sensations and passions felt by the original historical actors. Fourthly, they exercised a powerfully destabilizing effect on a range of existing Georgian entertainments, cultural forms and aesthetic theories, all of which were threatened, disrupted or overwhelmed by the burgeoning commercial appeal of this new mode of visual realism.

I

Thirty-one-year-old Philippe de Loutherbourg had already acquired a stellar academic reputation in Paris as a landscape and marine painter before he migrated to London in 1771 in order to escape a troubled marriage. Here he applied for work with the celebrity manager-actor of Drury Lane Theatre David Garrick who was gradually introducing a more naturalistic visual style of acting and staging to British theatre.[3] Ever since a visit to Paris in the mid-1760s, Garrick had wanted to revitalize Drury Lane's dull scenography along the spectacular lines associated with Giovanni Servandoni at the Paris Opera. Impressed by Loutherbourg's portfolio of landscape paintings and model maquettes, Garrick appointed him in 1772 on a large salary to take complete charge of Drury Lane's scene designs and spectacles.

Loutherbourg's scenographic speciality became the adult pantomime, the most lucrative popular staple of the British patent theatres. Pantomimes or Harlequinades, with their simple formula of a hero's chase after an abducted lover, had always glossed over nuances of plot and acting in favour of songs, jokes, magic tricks and spectacular effects. Even so, Loutherbourg's impact was paradigmatic: he immediately began to introduce a series of vivid naturalistic innovations, including concealed lighting and coloured filters, illuminated silk transparencies, moveable screens and backdrops, topographically accurate sets, and a new depth of perspective through reorganized flats, wings and side slips. Garrick – recalling his own successful restaging of a glittering Shakespeare Jubilee Parade at Stratford in 1769 – seems to have suggested that his new scenographer design another such real-life pageant scene to give fresh zest to their rather tired Drury Lane pantomime repertoire.[4]

Over the next few years Philippe produced simulated mechanical reenactments of three social pageants for insertion in Garrick's pantomimes. Each real-life pageant had already proved popular with the newspaper-reading public of the previous summer season. In October 1773 he used *The Masque of Alfred*, a stock farce about the mythic Saxon origins of the English navy, as a vehicle for inserting a meticulous new reenactment of George III's Spithead naval review of the previous year when the British fleet had mounted a five-day display of sail power in front of wildly enthusiastic crowds. A year later, *The Maid of Oaks*, a new Drury Lane comedy by Colonel John Burgoyne, again featured a glamorous festivity of the previous season. To celebrate his engagement, Burgoyne's nephew, Lord Stanley, had organized an extravagant pastoral masquerade at his Surrey mansion, with a guest list so lustrous, it forced the temporary closure of the House of Commons. Robert Adam designed and built an elaborate outdoor pavilion through which the guests paraded in masque. Finally, in what was clearly becoming a seasonal formula, the feeble 1775 Harlequinade, *Queen Mab* carried a Loutherbourg simulation of a lavish Venetian-style regatta held on the Thames in 1773. Although the regatta had been a public event, it had caused widespread popular frustration because tickets to the viewing enclosures were exclusive, and ordinary spectators on the banks of the river had been unable to see because of the rows of aristocratic pleasure craft moored along the banks.[5]

Perhaps the most striking feature of all three of Loutherbourg's simulations was their rapturous reception, irrespective of the feeble quality of the host play or the arbitrariness of the theatrical context. Theatre critics and spectators appeared to be dazzled by the technical and historical accuracy and the naturalistic realism of this new form of spectacle. When preparing the Spithead naval review, Loutherbourg hired the former sailor and marine painter, Dominic Serres the Elder, to build accurate model ships and to advise on appropriate sail settings.[6] The ships sailed among waves so shimmeringly translucent that observers wondered if the water was real. One reviewer described the sail-past as 'so critically exact that I can hardly give human invention credit for the execution ... in point of distance, perspective, etc, it is not chargeable with the slightest impropriety'. 'Numberless and different kinds of vessels were beheld under weigh, with their sails full, making the different tacks', wrote another.[7] The same critics panned a 'glaringly absurd' Covent Garden imitation, where 'one ship had sails aback yet sailed at twelve knots in the wind's eye'.[8] Philippe's simulation of Stanley's colonnaded pavilion was praised similarly for exactly replicating Adams's own published plans. The Venetian regatta scene went one better by introducing a startling Loutherbourg innovation of having mechanical puppets row the barges, 'moving', according to the *General Evening Post* reporter, 'so critically exact to the music'.[9] Since the puppets had been designed by Philippe's close friends John Joseph Merlin and Henry Jacquet-Droz, two of the finest automata builders in Europe,[10] it was not surprising to find the regatta hailed by the same newspaper as 'one of the completest pieces of mechanism seen on a Theatre'.[11]

The disproportionate success of these spectacles, along with the heavy investment in wages, materials and time to prepare them, accentuated Garrick's temptation to recycle, often in increasingly arbitrary contexts. Loutherbourg's designs for Stanley's *fête champêtre* cost a massive £1500, the most ever spent on scenery at Drury Lane. Yet it was worth it: the play fetched a profit by running for 25 nights in the first season and then being many times revived. Despite the marvel of Loutherbourg's fleet simulation, *The Masque of Alfred* was panned, but it retrieved its losses when Garrick shrewdly transplanted the naval review into a more lively stock comedy, *The Fair Quaker of Deal*, changing the name to *The Fair Quaker of the Humors of the Navy*. The new pantomime flourished, notwithstanding the lugubrious title and murmurs in the press about Garrick's money-grubbing. [12]

A few critics did wonder how Loutherbourg had come to be armed with his pencil at Lord Stanley's exclusive private festivities well in advance of Colonel Burgoyne's Drury Lane play on the subject. Given that Burgoyne was the host's uncle and a close personal friend of Garrick, it looked as if Stanley had been persuaded to mount the festivities so that Drury Lane could re-enact them for profit.[13] Still, Garrick was not the only one to recognize and exploit the commercial appeal of the new genre. Loutherbourg's reenactments generated a flurry of imitations in London and provincial theatres. The Spithead naval review was still being replicated a decade after the original: it surfaced in multiple forms, including as a Royal Academy painting, a

Marylebone puppet theatre show, and as one of the earliest scenes of Robert Barker's commercial panorama.[14]

II

David Garrick might have provided the initial impetus for Philippe's pageant simulations, but it was under the successor management of Richard Sheridan from 1776 that he developed the genre to a new level of aesthetic ambition and audience popularity, leading to its permanent instantiation within London's entertainment world. The catalyst was less the advent of Sheridan's playwright genius than Britain's involvement in the American War. Patriotic excitement, intermixed with growing popular anxiety after France and Spain joined the rebellious American colonists, created an ideal climate for introducing a new type of theatrical reenactment. Then as now, war generated unusually large and susceptible audiences eager for news from the front. Always commercially shrewd, Philippe sensed the opportunity to meet this need by reproducing spectacular 'real life' narratives of heroism and suffering. Within two years of the outbreak of war, Sheridan commissioned Philippe to produce a new type of military and naval reenactment designed to eclipse the earlier social pageants in technological complexity and audience affect.

The Camp, a two-act afterpiece, was written by Sheridan in 1778 specifically in order to display Philippe's mechanical re-creation of a military camp at Coxheath near Maidstone, 40 miles from London. Recent losses in America, combined with France's entry into the war, had whipped up a flurry of invasion rumours. Sheridan hardly needed the *Morning Chronicle* to remind him that 'when we are afraid of an enemy, the sight of men in arms who are to protect us, is of all spectacles the most agreeable'.[15] Lord North's Tory Ministry also took the point: they quickly set up a chain of camps along the South East coast with the combined purpose of resisting any French invasion, giving national reassurance, and enabling raw troops to practice field manoeuvres. Coxheath, the largest camp, mushroomed into a tented city housing 17,000 troops and civilians. 'Camp madness', as Sheridan later called it, became the fashion of the social season, making Coxheath a magnet for the *beau monde*, as well as for streams of casual tourists.[16]

Philippe spent several weeks living at Coxheath during which he sketched soldier's uniforms and weaponry with such detailed accuracy that they have become a staple source for modern military re-enactors.[17] His *piece de resistance* was to design and build a 'Lilliputian corps' of military automata, representing 18 infantry regiments marching and drilling in exact time to martial music. With pardonable exaggeration, the *Town and Country Magazine* called it 'one of the grandest and most warlike scenes that has ever been exhibited in this country ... and [which] must make a proper impression on our enemies'.[18]

Capitalizing on *The Camp's* opening season success of 56 performances, Sheridan grabbed the opportunity in November the following year to milk further patriotic sentiment and satirize the inertia of Lord North's Tory Ministry by commissioning a pointed historical reenactment for insertion in a play called *The Critic*, which he wrote himself. Philippe obliged by recreating Drake's historical sea victory over the Spanish Armada, with model ships tossing so realistically on spindle-moved waves that they aroused widespread 'astonishment'. It was inserted as the final scene of the wittily self-reflexive spoof about theatre management. As well as earning the satisfying sum of nearly £11,000, Sheridan enjoyed the further piquancy of attributing the Armada idea to the Tory-supporting King who'd called for the nation to emulate Drake's spirit.[19]

One of the rare British victories of the war occasioned Philippe's subsequent martial spectacle, a storming of the Spanish Fort of Omoa at the bay of Honduras in October 1779. Within a month of the military dispatches reaching England, Philippe produced three scenes for insertion in a pantomime called *Harlequin Fortunatus*. His reenactment had British sailors swarming up ladders on the sheer sides of the fort, being hit by rifle fire and plunging into the sea, until they eventually forced the Spanish governor to surrender. As in *The Camp*, mechanical soldiers and live actors were combined in a hybrid mix intended to convey both immediacy of action and an epic scale. The *Public Advertizer* surmised that, 'There certainly never was a more difficult Exhibition attempted in a Theatre, than to represent the Attack, Rout, and Confusion of a Fort stormed by a desperate Enemy: But here by the force of Perspective, the Eye commands the Operations both by Land and by Sea'. The reporter could hardly believe that Loutherbourg had somehow 'methodized this confusion'.[20]

Impelled by an ever more intense drive to replicate the full psychic vectors of social experience, these martial spectacles shifted the reenactment genre from celebrations of community pageantry to portrayals of 'extreme history'. In terms of the fashionable Burkean aesthetic, 'beautiful' and 'picturesque' real-life scenes gave way to 'sublime' instances of war heroism and suffering. The *Ladies Magazine* was thrilled, for example, that Loutherbourg's *Storming of Fort Omoa* 'represented the singular instance of an English seaman's bravery, recited by Captain Dalrymple's dispatch ... a Spaniard found without arms being furnished with a cutlass by a sailor who afterwards conquers him, then spares his life'.[21]

Several critics observed, however, that the introduction of real-life war scenes was damaging the traditional structure and tone of the Harlequinade. The *Morning Post* commented acidly in October 1778 that plots had become a mere 'jeu d'esprit', or 'flimsy story' for the purpose of introducing Mr de Loutherbourg's reenactments 'with a kind of dramatic propriety'.[22] Scenes of death and sacrifice also seemed tasteless and incongruent when set beside 'the flippant personage of Harlequin.'[23] Other reviewers welcomed Loutherbourg's introduction of a new 'spirit of military ardour' to the popular theatre, but they were troubled by the element of commercial calculation involved. It was rumoured that Sheridan and his scenographer were preparing martial simulations in advance of battles to capitalize on 'the first achievement that should arise'.[24] The contemporary French theorist Jean Baudrillard would regard this as an early instance of a media simulacrum – the reenactment of a real-life event that has not actually taken place. To eighteenth-century critics it seemed rather that the stage, always the mirror of the age, had become a literal battlefield.

Reviewers were also quick to discern that these spectacular reenactments had instigated a marked change in the traditional roles and hierarchies of the Augustan theatre. The recent cult of celebrity that had grown up around actors like David Garrick and Sarah Siddons was now extending to those stage personnel once regarded as little more than artisans. Many critics noted that Philippe, the creator of special effects, had become the real star of Drury Lane productions. The *Saint James Chronicle* stated bluntly that: 'the writer and the composer are so totally eclipsed by the Painter that the Entertainment of *The Camp*, will always be attributed to the Talents of Mr de Loutherbourgh'.[25] This point was not lost on Sheridan, who both satirized and fostered his scene-designer's celebrity by writing Philippe into several of his wittily meta-theatrical playscripts. *The Camp*, for example, featured 'Mr Lanternberg the great painter' whose assistant, a stock Irishman called O'Daub, had just finished making real-life sketches of Stanley's *fete champetre* and was now repeating the formula for a new play about a military camp. *The Critic* similarly notes that *Storming of Fort Omoa* was yet another example of 'the miraculous powers of Mr de Loutherbourg's pencil', but also gestured wryly at Sheridan's own directorial cynicism in using martial reenactments within his plays.[26] *The Critic*'s Mr Puff, a stage director, while rehearsing a play explains his arbitrary inclusion of Loutherbourg's Spanish Armada simulation with the words, 'Now, gentlemen, this scene goes entirely for what we call SITUATION or STAGE EFFECT, by which the greatest applause may be obtained, without the assistance of language, sentiment, or character'.[27]

A few critics also felt that Loutherbourg's martial reenactments had pushed the principle of realism in the theatre to paradigmatic new levels.[28] It was commonplace for newspapers to say of his reenactments that they 'appear as in real life' and produce 'unaccountable deceptions', but spectators were also sometimes explicitly recommended to substitute his theatrical simulations for the original experience. [29] In *The Camp*, Sheridan had the scene painter O'Daub tout his mechanical and painted reenactment to the audience in preference to the cost and effort of making a real-life visit to Coxheath.[30] Evidently some viewers took the advice to heart: George III was reported to have delayed a scheduled review of the troops at Coxheath because he'd already seen Philippe's version of the event.[31]

Conflations of reality and theatrical illusion could have unexpectedly dangerous consequences. Colonel Burgoyne, amateur playwright and original conniver of Loutherbourg's *fête champêtre* reenactment of 1773, became Commander of the British forces in the US where he continued to write for and attend theatre performances. On one occasion in 1776, however, he failed to realize that real-life Yankee rebels had launched an attack on his forces because he confused it with a theatrical farce he was watching called *The Blockade of Boston*.[32] Spectators could also become victims of excessive realism. During an early performance of Loutherbourg's *Storming of Fort Omoa* the audience re-experienced the original sufferings of the troops all too literally: the Drury Lane stage became so enveloped in acrid gunpowder smoke and overwhelmed by the din of detonations that no-one could see or hear.[33]

Theatrical simulations also began to influence Philippe's practice as a salon artist. Though he had occasionally adapted oil painting scenes for use in theatre sets, the reverse occurred in the aftermath of *The Camp*. It was the fame of the Coxheath simulation that brought him lucrative new commissions to paint portraits of an adjacent military camp at Warley. Surviving notebooks show his pains to attain a comparable level of visual realism, something that threatened to endanger his simultaneous efforts to establish himself as an elite Academic painter in Britain. Sir Joshua Reynolds, President of the Royal Academy and author of the famous *Discourses*, repeatedly distinguished between the artisan-like mechanical skill of copying nature's details exactly and the true artist's imaginative and mental task of embellishing nature to produce a single unified and exalted idea. The former aesthetic appealed to the untutored popular sensibilities of the middle classes who flocked to London's commercial entertainments, the latter was the province of aristocratic connoisseurs, art patrons and expert collectors. A published review of Loutherbourg's 1780 Royal Academy painting, *The Troops at Warley-Camp reviewed by His Majesty*, came dangerously close to praising the work as a realistic reenactment rather than a work of the idealizing imagination. The critic marvelled at the intricacy of detail: 'it is so full of business ... it is hardly possible to pay attention to the whole. We look for miles into the picture; His Majesty, attended by his Aids-de-Camp, Equerries, etc is seen in the centre of the piece, while the Camp and Line extends from right to left'.[34]

Loutherbourg's martial reenactments, even more than his earlier pageants generated scores of imitators. Within a year of *The Camp's* London debut, infantry simulations modelled on Loutherbourg's Coxheath originals were being staged in the major provincial theatres of Manchester, Bristol and Dublin.[35] By 1800 at least 16 plays incorporating similar martial spectacles had appeared in the London patent theatres.[36] Well before this, they were being referred to in the press as a distinct new entertainment genre known as 'military and naval spectacles' which often transcended their origins within theatrical performances. Elaborate mock battles with realistic settings, models and armaments were staged for fee-paying audiences outside famous public houses like Thomas Keyse's Waterman's Arms in Bermondsey with its specially-built three acre battle site, or in Astley's Circus hippodrome which could be flooded to re-enact miniature naval engagements known as naumachia.[37]

In the immediate aftermath of *The Camp*, Philippe pioneered another new form of theatrical simulation centred on real-life tourist travels. The handsome sum of £160 that Sheridan had paid him to study Coxheath also funded a sketching tour of Derbyshire in the same year. Both he and Sheridan had long observed the burgeoning popular interest in visiting, sketching or reading about tours of the wilder northern regions of the British countryside once regarded with aesthetic indifference or disdain. Unable to afford the costs and unwilling to take the risks of the traditional aristocratic Grand Tour of Europe, the affluent middle classes were increasingly drawn to the delights of armchair travel, including the remoter, more exotic regions of their own country. Beginning in mid century, travel books by the likes of Thomas Gray, Thomas Amory, Thomas West, William Hutchinson and William Gilpin had inaugurated a vogue for the picturesque and sublime beauties of the Lake District. Though relatively little had been written about Derbyshire which Gilpin regarded as rather too savage an environment,[38] William Bray changed this with his influential *Sketch of a Tour into Derbyshire and Yorkshire* published in 1777. The following year the *London Magazine* contained an even more fulsome follow-up article describing the seven great natural wonders of Derbyshire. Reading this seems to have inspired Philippe with the idea of undertaking his own exploration and sketching of the district, which could then become the basis for an elaborate series of special effect scenes around which Sheridan could construct a new pantomime. The result, an exceptionally slight Harlequinade called *The Wonders of Derbyshire* with 12 Loutherbourg scenes was a smash hit: it ran for 49 nights and attracted breathless comparisons with the theatre scenography of Giovanni Servandoni and the landscape art of Claude Lorraine.[39]

The disruptive trends inaugurated by Philippe's earlier forms of reenactment were, if anything, exacerbated by this new venture into virtual tourism. *The Wonders* required an even heavier investment of time and money, taking seven months of feverish work to prepare from sketch to production, forcing the management in the meantime to rely on stock plays.[40] The naturalistic brilliance of the new sets also completely overwhelmed the plot, derided by the *Westminster Magazine* as 'absolutely the most contemptible'.[41] Harlequin and Columbine did little more than chase along Philippe's path. Critics rightly identified him as the true 'author'; while the play's true subject matter was the scenery of mountain, cave and valley that he had recreated.

As if to flaunt this departure from pantomimic orthodoxy, a printed guide discussed the real-life veracity of the topography, geology and folk lore of each Loutherbourg scene and was offered for sale to patrons of the play.[42] One critic acknowledged the verisimilitude but wished that the artist had also taken along 'a man of some dramatic genius, or at least of Talents for the Invention of a Pantomime'.[43] Though appreciative of the naturalistic beauty of the sets, several reviewers mourned the disappearance of traditional pantomimic 'tricks', 'laughable incidents' and 'genius of buffoonery'.[44] Some thought that in following the aesthetic of the sublime popularized by Edmund Burke, Philippe had 'lent too much to the frightful',[45] an aesthetic unsuited to those 'for whom pantomimes are principally intended'.[46] Moreover, as with his martial spectacles, Philippe's success made topographical reenactment a permanent feature of the London and provincial stage repertoire, not least in his own 1780's travelogue blockbusters of the South Seas

Robinson Crusoe and *Omai, or a Trip Round the World*.[47] Of the latter, *The Times* commented: 'To the rational mind, what can be more entertaining than ... to bring into living action, the customs and manners of distant nations. To see exact representations of their buildings, marine vessels, arms, manufacturers, sacrifices and dresses'.[48] By the 1780s, the age of reality reenactment had emphatically arrived.

III

This point was given potent confirmation when in January 1781, Philippe launched a unique new form of visual commercial entertainment, a hybrid of theatre and kinetic art which he called the *Eidophusikon*. On entering a specially-built miniature theatre located inside his Leicester Square house, viewers paid five shillings to watch a type of giant peep show, 8ft by 6ft, within which was projected a programme of five landscape scenes of nature in action. Though many of the *Eidophusikon*'s illusionist effects derived from Philippe's prior scenographic and artistic experience, they had been transformed into a stunning new visual medium by his ability to achieve precise concentrations of lighting and shadow, his use of coloured filters, magic lantern slides, three-dimensional model topographies, intricately-built clockwork automata, back- and front-lit transparencies, and spindle-driven back-screens. In short, Philippe de Loutherbourg had invented a primitive form of cinema.

At first, the *Eidophusikon* carried no 'reenactment' segment, probably because Philippe, anticipating election as a full Fellow of the Royal Academy of Art, took pains to purge his show of any of the more vulgar associations of either theatre or commercial spectacle. His high entry fee and decorous furnishings trumpeted gentility: the show's repertoire was designed to be seen as a refined extension of his professional practice in picturesque landscape art. His motivation for developing it, he told a journalist of the *European Magazine*, had simply been 'to add motion to resemblance' because 'the most exquisite painting represented only one moment of time in action'.[49] All the initial *Eidophusikon* scenes – dawn over the Thames, noon at Tangier, sunset on the Bay of Naples, moonrise on the Mediterranean and a shipwreck during a storm at sea – were derived from former easel art exhibits. However, once he had successfully persuaded critics and fellow artists to hail the show as 'a new species of painting', he felt emboldened to unleash the drawcard of reenactment. Before the first season was over, he replaced the rather lacklustre Tangier scene with a patriotic martial segment showing French and Dutch prizes being towed into the Port of Plymouth.[50] In effect, it was the first war movie.

Although the *Eidophusikon's* successive programmes evolved in a variety of interesting ways, it was not until Philippe's last direct involvement with the show in early 1786 that he added to its reenactment repertoire. By this time he no longer actually owned the show. Realizing he could no longer satisfy the London crowd's voracious demand for novelty without seriously damaging his fine art pretensions, he had sold the *Eidophusikon* to a former employee, John Chapman. The latter was busy preparing a new, more populist, programme when on 6 January 1786 a spectacular marine disaster called out for Loutherbourg's re-enactive genius. An East-Indiaman, the *Halsewell*, had been wrecked off the Dorset Coast in circumstances that created a media sensation. Within three weeks, advertisements boasted that Loutherbourg had produced 'an exact, awful, and tremendous Representation of that Lamentable event'. [51] A printed version of the story taken from the real-life account of a survivor was also offered for sale as a supplement to the performance.

It was a harrowing story. The proud new Indiaman had been gradually pulverized by the ferocity of the storm until it struck on the steep, jagged cliffs off Purbeck Island. Hundreds of sailors threw themselves into the water or slipped off a narrow rock ledge to be drowned or pounded to pieces on the rocks. The truly heart-rending dimension of the disaster, however, centred on the presence aboard of seven young women 'distinguished for their beauty and their accomplishments'.[52] Supposedly destined for romance in India, they drowned with 'terrible shrieks' when the for'ard roundhouse broke off and sank. In the meantime the Captain Richard Pierce, whose two daughters were among the group, had gathered all the girls in his arms. 'Struggling to suppress the paternal tear which then burst into eye', he waited for the waves to engulf them all.[53] Notwithstanding the efforts of local quarrymen to rescue the survivors from the cliffs, only 74 out of 240 crew and passengers survived.[54]

We don't know exactly what the *Halsewell* reenactment entailed, though it was probably adapted in part from earlier shipwreck scenes that Philippe had used in various visual spectacles. An artist, W. H. Pyne, who viewed the new reenactment, claimed that Loutherbourg had developed a new 'picturesque of sound', reproducing the 'horrors of wind, hail, thunder, lightning, and the roaring of waves ... with such marvellous imitation of nature, that mariners have declared ... that it amounted to reality'.[55] The torrential storm and lethal cliffs evoked the voguish aesthetic of the sublime, but it was the sentimental spectacle of heart-rending pain and suffering that ingrained itself ever after on Pyne's memory. So vivid was the experience that he felt himself an actual participant in the tragedy, feeling personally devastated by the futility of firing a distress signal into the teeth of the 'raging tempest'. 'I ... could not listen to, even in this mimic scene, without being reminded of the heart-sickening answer which sympathetic danger had poured forth ... a hoarse sound to the howling wind, that proclaimed, "I, too, holy Heaven! Need that succour I fain would lend'. [56] One could hardly find a better example of what Jonathan Lamb has called '[reenactment's] close spectatorial relationship to scenes of pain'. *Titanic*, weep your heart out.

V

Ironically it was only Philippe's last lucrative venture into reenactment that led him to appreciate some of its aesthetic and social drawbacks. Following France's declaration of war on Britain on 1 February 1793, popular patriotic fascination with the realistic details of conflict flared once again. This time, however, Philippe capitalized on the trend by producing oil-paint reenactments rather than theatrical or movie simulations. News in late July that the Duke of York, commander of the British forces on the Continent, was besieging the French army at Valenciennes in Belgium was received with wholesale British rejoicing. A consortium of English and Swiss engraver-publishers immediately commissioned Philippe to travel to the battlefield to prepare a large panoramic canvas that could be exhibited in public and engraved for profit. The *Morning Post* wondered sarcastically whether the siege was to be performed again 'for the amusement and instruction of [Loutherbourg]'.[57]

Ignoring such barbs, Philippe set off on 30 August, accompanied by James Gillray whom he hired to sketch portraits of the soldiers while he would concentrate on the landscapes of battle. After braving storms, sea-sickness, a bout of flu and reports of French atrocities, they rode to the battlefields in an old English chariot along roads lined with fresh crosses. Picking a point on the outskirts of Valenciennes, they began sketching, ignoring the bullets that occasionally whizzed overhead.[58] After a month's tour, they realized that the British had somehow managed to snatch defeat from the jaws of victory. Even so, on returning home they were eagerly welcomed by news-hungry Britons, including the King, who summoned them for a preliminary personal viewing. Fortunately for Philippe, George found the sketches of the shattered landscape persuasive, but the portraitist Gillray had to endure the King's scorn because the faces of the English officers were insufficiently realistic. Such are the hazards of literalism.

Still, the pair reaped a financial bonanza: Philippe's massive 6ft x 8ft panoramic painting was shown at the Historic Gallery in Pall Mall for a 1s a head, and engravings sold for 3 guineas each. He earned the colossal sum of £500 plus Gillray's expenses. [59] Significantly their chief commercial rival, battle painter Mather Brown, could only try to counter their reenactment advantage by claiming he had experienced even more authentic siege bombardments while fighting in the American War.[60] Philippe had again inaugurated a trend; by the end of century it was mandatory for war artists to visit battle-sites, tour warships, interview participants and utilize formal battle plans. [61]

For Philippe and Gillray a second lucrative commission followed in 1794, this time to paint the first great British sea victory of the war, the 'Glorious First of June', when Admiral Howe defeated a large French fleet off Brittany. Though they obviously couldn't tour the battle-site, they did sketch the fleet at Portsmouth and interview numbers of the officers and sailors. For once, however, Loutherbourg's wish to display fine art principles – unity of theme and composition, dramatic focus and an idealizing set of moral principles – overcame his commitment to literalness. When the massive painting was exhibited, its depiction of the duel between the opposing British and French flagships was sharply criticized by Admiral Howe for inaccuracy. The battle hero, Fleet Master James Bowen, also berated it as an implicit slur on the courage and skill of Howe's sailors. Loutherbourg had moved the position of the French flagship for better composition, but this seriously distorted the strategic alignments and outcomes of the actual battle.

Philippe had been hoist on his own petard. Though fellow artists praised the painting for its depiction of exemplary heroism and compassion, the literalist criticism of the naval participants damaged his credibility. He found himself caught in an increasing tension between those artists committed to fine-art, idealizing principles in the vein of Sir Joshua Reynolds, and those prepared to make ready money with eager middle-class audiences by producing hyper-realistic reenactments. The former held that the artist's mission was to deploy allegorical references and moral principles within classical tableaux; the latter wanted to replicate all the literal details of blood, pain and chaos of warfare. Both Augustans like Reynolds and Romantics like Coleridge saw this distinction as one arising between the proper imaginative imitation practiced by the true artist and the poet and the mechanical copying or vulgar mimesis executed by the mere journeyman or commercial entertainer.[62] Ultimately this aesthetic divergence led to the expulsion of battle reenactments from within the formal hierarchy of British fine art painting. Henceforth academic history painters concentrated on producing idealized and Romantic versions of dramatic episodes such as Nelson's death, while realistic battle scenes became the province of professional specialists recruited from within the armed services.[63]

Philippe's *Victory over the French at Maida* in 1806 proved to be the last major battle painting of the Napoleonic wars. Sadly, it was not a success. Forced to choose between his Academic status and the commercial profitability of the reenactment form, Philippe hovered uncertainly between the two conflicting imperatives.[64] The painting provided a sobering finale to his career as a reenactment expert and fine-arts easel painter. Perhaps he would have been consoled to know that 200 years later many history and documentary film-makers are faced with a similar tension between the odious popularity of commercial reenactment and the respectable austerity of academic art.

Notes

1. 'Queen Elizabeth attends Battle of Trafalgar Commemoration', available at http://the internetforum.co.uk/rf/Trafalgar.html.
2. See J. Lamb (2009) 'Historical Re-enactment, Extremity and Passion', *Eighteenth Century: Theory and Interpretation*, 49:3, pp. 239–250.
3. G. D'Arcy Wood (2001) *The Shock of the Real: Romanticism and Visual Culture, 1760–1860* (New York: Palgrave Macmillan), pp. 17–42.
4. By far the best analysis of Garrick's and Loutherbourg's stage innovations is C. Baugh (1987) 'Philip James de Loutherbourg and the Early Pictorial Theatre: Some Aspects of its Cultural Context', in J. Redmond (ed.) *The Theatrical Space* (Cambridge; New York: Cambridge University Press), pp. 99–128.
5. R. G. Allen (1965) 'Topical Scenes for Pantomime', *Education Theatre Journal*, 17:4, pp. 289–95.
6. For Dominic Serres the Elder, see P. H. Highfill Jr, A. Kalman Burnim and E. A. Langhans (1991) *A Biographical Dictionary of Actors, Actresses, Musicians, Dancers, Managers & Other Stage Personnel in London, 1660–1800*, vol. 13 (Carbondale: Southern Illinois University Press), pp. 256–7.
7. Allen, 'Topical Scenes', p. 291.
8. R. Joppien (1972) 'Die Szenenbilder Philippe Jacques de Loutherbourgs: Eine Untersuchung zu ihrer Stellung zwischen Malerei und Theater' (Ph.D. dissertation, University of Cologne), p. 53.
9. Allen, 'Topical Scenes', p. 295.
10. For Merlin, see A. French, M. Wright and F. Palmer (1985) *John Joseph Merlin: The Ingenious Mechanick* (Kenwood: Iveagh Bequest, Greater London Council); for Jacquet-Droz, see G. Wood (2002) *Living Dolls. A Magical History*

of the Quest for Mechanical Life (London: Faber), pp. xii–xvi; P. Metzner (1998) *Crescendo of the Virtuoso. Spectacles, Skill, and Self-Promotion in Paris during the Age of Revolution* (Berkeley and Los Angeles: University of California Press), pp. 167–177.
11. Allen, 'Topical Scenes', p. 295.
12. Joppien, 'Die Szenenbilder Philippe Jacques de Loutherbourgs', p. 52.
13. C. Dibdin (1803) *The Professional Life of Charles Dibdin*, 4 vols (London: The Author), pp. 73–5; Allen, 'Topical Scenes', p. 293.
14. Joppien, 'Die Szenenbilder Philippe Jacques de Loutherbourgs', pp. 53–4.
15. *Morning Chronicle* (3 August 1778) cited in G. Russell (1995) *The Theatres of War: Performance, Politics, and Society, 1793–1815* (Oxford: Oxford University Press), p. 36.
16. For the play, see C. Price (ed.) (1973) *The Dramatic Works of Richard Sheridan*, vol. 2, (Oxford: Clarendon), pp. 707–50; on camp culture and politics see, Russell, *Theatres of War*, pp. 26–51; R. W. Jones (Winter 2002) 'Sheridan and the Theatre of Patriotism: Staging Dissent During the War for America', *Eighteenth-Century Life*, 26, pp. 24–45; S. Conway (1997) 'Locality, Metropolis and Nation: The Impact of the Military Camps in England during the American War', *History*, 82, pp. 547–62.
17. M. Tully, 'Reenacting is Authentic!', available at http://www.nwta.com/couriers/8-96/reenacting.html.
18. Allen, 'Topical Scenes', p. 296.
19. J. Loftis (1976) *Sheridan and the Drama of Georgian England* (Oxford: Blackwell), pp. 120–23; Conway, 'Locality, Metropolis and Nation', p. 619.
20. Allen, 'Topical Scenes', p. 300.
21. Ibid., p. 299.
22. *Morning Post* (16 October 1778), cited in Joppien, 'Die Szenenbilder Philippe Jacques de Loutherbourgs', p. 200.
23. *Saint James Chronicle*, cited in Allen, 'Topical Spectacles', p. 298.
24. Joppien, 'Die Szenenbilder Philippe Jacques de Loutherbourgs', p. 200.
25. Price, *The Dramatic Works of Richard Sheridan*, vol. 2, p. 712.
26. Highfill et al., *A Biographical Dictionary*, vol. 4, p. 308.
27. Price, *The Dramatic Works of Richard Sheridan*, vol. 2, p. 711.
28. As early as 1776, the *Morning Chronicle* praised Philippe as 'the first artist who showed to our theatre directors that by a just disposition of light and shade, and critical preservation of perspective, the eye of the spectator might be so effectually deceived in a playhouse as to be induced to take the product of art for real nature'. See (17 February 1776) *Morning Chronicle*, cited in Highfill (1978) *A Biographical Dictionary*, vol. 4, p. 305.
29. *Whitehall Evening Post* (17–20 October 1778), cited in Joppien, 'Die Szenenbilder Philippe Jacques de Loutherbourgs', p. 204.
30. Joppien, 'Die Szenenbilder Philippe Jacques de Loutherbourgs', p. 200.
31. Ibid., p. 202.
32. Russell, *Theatres of War*, p. 27.
33. Allen, 'Topical Scenes', p. 300.
34. By An Artist (1780) *A Candid Review of the Exhibition (Being the Twelfth) of the Royal Academy no. XV* (London: H. Reynell), p. 14.
35. Joppien, 'Die Szenenbilder Philippe Jacques de Loutherbourgs: Eine Untersuchung zu ihrer Stellung zwischen Malerei und Theater', p. 205.

36. Ibid., pp. 206–7.
37. R. D. Altick (1978) *The Shows of London* (Cambridge, MA: Bellknap Press), p. 97.
38. W. Bray (1777) *A Sketch of a Tour into Derbyshire and Yorkshire* (London).
39. R. G. Allen (1961) 'The Wonders of Derbyshire', *Theatre Survey*, 2, pp. 54–66; S. Daniels (1993) 'Joseph Wright and the Spectacle of Power', in *Fields of Vision: Landscape Imagery and National Identity in England and the United States* (Cambridge: Polity Press), p. 59. Loutherbourg's painted curtain backdrop of the thunderous falls at Matlock High Torr became a permanent feature at Drury Lane performances until well into the 1790s.
40. Joppien, 'Die Szenenbilder Philippe Jacques de Loutherbourgs', p. 209.
41. Allen, 'The Wonders of Derbyshire', p. 57.
42. Derbyshire Man (1779) *An Account of the Wonders of Derbyshire, as Introduced in the Pantomime Entertainment at the Theatre-Royal, Drury Lane, by a Derbyshire Man* (London: G. Bigg), pp. 8–23.
43. *Saint James Chronicle* cited in Allen, 'The Wonders of Derbyshire', p. 57.
44. Allen, 'The Wonders of Derbyshire', p. 57.
45. *Morning Chronicle* (9 January 1779), cited in Joppien, 'Die Szenenbilder Philippe Jacques de Loutherbourgs', p. 213.
46. Allen, 'The Wonders of Derbyshire', p. 57
47. Joppien, 'Die Szenenbilder Philippe Jacques de Loutherbourgs', p. 224; Allen, 'The Wonders of Derbyshire', p. 54.
48. I. McCalman (2001) 'Spectacles of Knowledge: Omai as Ethnographic Travelogue', in M. Hetherington (ed.) *Cook and Omai. The Cult of the South Seas* (Canberra: National Library of Australia), p. 11; see also N. J. D. Hazelton (2002) 'Green to the Very Door: Painted Landscape on the Nineteenth-Century Stage', *Theatre History Studies*, 22, pp. 115–36.
49. Altick, *The Shows of London*, p. 121.
50. Ibid.
51. Ibid., p.125.
52. H. Meriton (1786) *An Interesting and Authentic Account of the Loss of the Halsewell. East-India-Man, with All Its Dreadful Circumstances* (London: W. Bailey), p. 9. See also T. S. R. Boase (1959) 'Shipwrecks in English Romantic Painting', *Journal of the Warburg and Courtauld Institutes*, 22, pp. 336–7.
53. (1786) 'East-Indiaman, Appendix to the Chronicle' in *Annual Register of World Events* (London: Longman), p. 227.
54. Meriton, *An Interesting and Authentic Account*, p. 37.
55. W. H. Pyne (1824) *Wine and Walnuts; Or, After Dinner Chit-Chat by Ephraim Hardcastle, Citizen and Dry-Salter*, vol. 1, 2nd edn (London), p. 292.
56. Ibid., p. 296.
57. D. Hill (1965) *Mr. Gillray, the Caricaturist: A Biography* (London: Phaidon). This contains a transcription of Gillray's journal which remains in the author's private possession, see p. 49.
58. Ibid., pp. 49–53.
59. The details of the contract between Loutherbourg and Messrs V. and R. Green and Christian de Mechel are found in the appendix to Anthony Griffith's article. See A. Griffiths (December 2003) 'The Contract for the Grand Attack on Valenciennes', *Print Quarterly*, 20:4, pp. 378–9.

60. P. Harrington (1993) *British Artists and War: The Face of Battle in Paintings and Prints, 1700–1914* (London: Greenhill Books; Pennsylvania: Stackpole Books in association with Brown University Library, Rhode Island), p. 70.
61. Ibid., p. 79.
62. Wood, *The Shock of the Real*, pp. 66–97.
63. Harrington, *British Artists and War*, p. 90.
64. Ibid., pp. 86–90.

Index